Against Method

Fourth Edition

Paul Feyerabend

VERSO
London • New York

First published by New Left Books 1975
Revised edition published by Verso 1988
Third edition published by Verso 1993
This edition published by Verso 2010
© Paul Feyerabend 1975, 1988, 1993, 2010
Introduction to the fourth edition © Ian Hacking

3 5 7 9 10 8 6 4

Verso
UK: 6 Meard Street, London W1F 0EG
US: 20 Jay Street, Suite 1010, Brooklyn, NY 11201
www.versobooks.com
Verso is the imprint of New Left Books

ISBN-13: 978-1-84467-442-8

British Library Cataloguing in Publication Data
A catalogue record for this book is available from the British Library

Library of Congress Cataloging-in-Publication Data
A catalog record for this book is available from the Library of Congress

Typeset by Hewer Text UK Ltd, Edinburgh
Printed in the US by Quad/Graphics Fairfield

Contents

Introduction to the Fourth Edition

by Ian Hacking

'*Against Method* is more than a book: it is an event.'[1] That was what it felt like, when the work came out in 1975. Feyerabend was notorious, adored by the young, loathed by the established. The turbulent sixties were winding down, and here was an intellectual testament to the ferment. This was the Woodstock of philosophy. The book should now be read in two ways, both as a part of that era, and also as a contribution to intellectual life in the long term.

There are many lovely things about the book. The first is the Analytical Index, 'Being a Sketch of the Main Argument'. This is not some machine-readable abstract of the type now required by scholarly journals. Paul Feyerabend is telling *you*, in his own plain (and thereby elegant) prose, what he thinks is interesting, chapter by chapter. Yes, it is OK if you skip a couple of chapters, or read the book from back to front. This is not to say you should not read the work sentence by sentence, but the great merit of a book is that you can take it hitch-hiking or to a sit-in, and read a bit while you are munching a few pilfered tomatoes or sheltering from a storm. You can pick up an idea, chase it, and relocate it in the Analytical Index, all the while being in a physical relation to the pages upon which you can scribble expostulations, if that is your wont.

I have been saying 'this book'. That is doubly wrong. First because, as Feyerabend truly said, '*AM* is not a book, it is a collage.'[2] Secondly, because there is more than one collage. There is the first edition of 1975, and the radically revised one of 1988, and then the third edition of 1993, reprinted here. The 1988 version is far more manifestly a collage than that of 1975, although the earlier printing was much more handsome. Feyerabend went on changing the text, but the biggest changes were for 1988. I shall mention some of them below. He is not quite right about the final changes he made, as described in the preface to the third edition below. Yes, the 1988 Chapter 20, not found in 1975, has been dropped. It is all about objectivity and

1. Jean Largeault, at the end of his review of the book in *Archives de Philosophie* 39 (1976), p. 389. This essay, by the most rigorous French philosopher of the sciences of his day, was far more perceptive than most of the English-language reviews at the time.

2. Paul Feyerabend, *Killing Time*, Chicago, 1995, p. 139.

the construction of objects, scientific and other. The 1988 Chapter 19 ('What's so great about science?') has been much extended. He says that he rewrote Chapter 16, which is essentially the 1975 Chapter 17, for the 1988 edition. That is misleading. He chiefly cut six pages from it: a discussion of incommensurability, a topic which, as he indicates in the Preface to the Third Edition, had been worked to death by 1993. And he added the epilogue on relativism, a bone on which he chewed over and over again.

The publishers of all three books are in effect the same, for New Left Books was the original trade name of Verso. Feyerabend had his little battles with New Left Books – see his amusing letters to Imre Lakatos, which include a frustrated cable from Feyerabend in New Zealand to Lakatos in London, dated 2 August 1972.[3] In the 1988 preface, Feyerabend indicated, in broad strokes, what he added, rearranged, or cut from 1975. I am by no means sure he was right to make the changes.

To Imre Lakatos

In 1975 there was a dedication, 'To IMRE LAKATOS Friend, and fellow-anarchist.' It was removed in 1988. Nothing odd about that – Lakatos had died suddenly in 1974. In 1975 there was a single moving paragraph explaining that the book had been intended to be published in tandem with an equally vigorous response by Lakatos, one which was never written. In 1988 Feyerabend put this thought into a longer but not more effective preface. Here he repeated the dedication in clumsy embedded prose: 'I therefore dedicate also this second, already much more lonely version of our common work to his memory.' The 1975 front matter – a dedication on an empty page, followed by a brief paragraph on another empty page – is far more moving, and no less intellectually telling. And so, in my opinion, it goes. *Of course* the publishers could not reprint the first edition rather than the final one. Happily they have now put the 1975 original online.

One fundamental difference is that in 1975 the book had a long Chapter 16 on Lakatos, the ironically named 'fellow-anarchist' of the subtitle. It was deleted and in 1988 the analytical summary of the chapter was reduced to a mere footnote to the preceding chapter.[4] It refers us to

---

3. *For and Against Method: Imre Lakatos and Paul Feyerabend*, edited by Matteo Motterlini, Chicago, 1999, p. 290.

4. Note 11, p. 161 below.

an adequate but passionless paper printed elsewhere.[5] So the 1988 book is no longer, as Feyerabend put it in 1975, 'a long and rather personal *letter* to Imre' – one such that 'every wicked phrase it contains was written in anticipation of an even more wicked reply from the recipient'.[6]

Lakatos and Popper

Feyerabend speaks for himself. An introduction by someone else is wanted only to suggest the historical setting in which the first book was published. The following notes are for people born after 1975, for whom the event called *Against Method* is somewhere back in prehistory, like the fall of the Berlin Wall. Since the book is dedicated to Imre Lakatos (1922–1974), we can start with their relationship, although that means I shall subsequently have to move backwards in time. For a deeper grasp of the friendship between Lakatos and Feyerabend, dip into *For and Against Method*.[7] It includes 250 pages of rambunctious, irreverent, but astute letters between the two men, written between 1967 and 1974, exactly the period when *Against Method* came into being.

Lakatos, educated in Budapest, had a turbulent youth. Towards the end of the Second World War he led a cell in the Communist resistance against Hungarian fascism. He became an influential figure in the party, was disgraced, jailed, released, and left Hungary after the failed rebellion of 1956. He arrived at the University of Cambridge. His PhD thesis was published as *Proofs and Refutations*. It is one of the most original twentieth-century contributions to the philosophy of mathematics, although it is usually regarded more as pedagogy than as philosophy. He turned to the philosophy of the sciences and gravitated, in 1960, to the London School of Economics, where Karl Popper ruled.

No philosopher of the sciences was more admired by working scientists than Popper. His watchword, or phrase, was 'Conjectures and Refutations'. Science is hypothetico-deductive. Scientists frame conjectures and test their logical consequences. A proposition is scientific if and only if it is falsifiable. Otherwise it is 'metaphysical' – not meaningless or useless, as logical

---

5. Paul Feyerabend, 'The Methodology of Scientific Research Programmes', in *Problems of Empiricism: Philosophical Papers II*, Cambridge, 1981, pp. 202–31.

6. Paul Feyerabend, *Against Method*, New Left Books, 1975, p. 7 (not numbered).

7. See note 3.

positivists tended to say, but in need of clarification, dialectical analysis, and deep thought, in order to be reworked into something testable. On numerous occasions Feyerabend was to recall that Popper began his class by saying that there is no scientific method. And then (said Feyerabend) he began to go wrong, enunciating, in effect, the method of conjectures and refutations.

In 1969, Lakatos inherited Popper's chair. He devised a 'Methodology of Scientific Research Programmes' which was, in Lakatos's own terminology, a progressive problem shift from Popper's inquiry into the nature of rationality and science. Lakatos continued the tradition of the 'Popper Seminar', a weekly happening during term. Under both Popper and Lakatos it was famous for confrontation. An invited guest seldom got through more than ten minutes of exposition before being subject to violent criticism. A great many people hated the experience and feared what they thought was hostility. Some loved the ambience. One was Feyerabend. I do not know when the two men, Lakatos and Feyerabend, first met; in the early 1960s, perhaps. By 1967, when the published correspondence between them begins, they were soulmates.

In 1965, Lakatos organized an important meeting in London, the first major collective response to Thomas Kuhn's *Structure of Scientific Revolutions* (1962). There were already plenty of conservative criticisms of Kuhn published or in the works. But the core of the meeting was radical criticism from a Popperian and 'post-Popperian' point of view. Lakatos thought that conference proceedings ought to be written after the conference, so that the authors learned from what had gone on. The boring papers of his own conference were published in standard channels, three forgotten volumes, but the main upshot was the memorable *Criticism and the Growth of Knowledge*.[8]

It includes pieces by Feyerabend, Kuhn, Popper, Toulmin – and Lakatos's first sustained statement of his new methodology, which constitutes almost half of the book. Feyerabend says in a note that his own essay for the volume, 'Consolations for the Specialist', was presented at the Popper seminar in 1967, but to judge by later letters to Lakatos, it was still being reworked in 1968. In fact it is not clear which work in progress becomes which final work. He speaks of 'my Kuhn paper'[9] and, six weeks later, 'my latest anti-Parmenidian

---

8. *Criticism and the Growth of Knowledge*, edited by Imre Lakatos and Alan Musgrave, Cambridge, 1970.

9. *For and Against Method*, p. 120.

paper,[10] both of which the editor of the correspondence identifies as 'Consolations'. A postscript to the first letter says he is thinking of calling his anti-Kuhn paper 'Against Method', adding parenthetically 'this in analogy to Susan Sontag's *Against Interpretation*'.[11] In the autobiography he refers to his 'pro/anti Kuhn paper'.[12] Certainly in 'Consolations' we get Feyerabend's most famous utterance, 'anything goes'.[13] But he published a paper with the title 'Against Method: Outline of an Anarchistic Theory of Knowledge', also in 1970.[14] This really is a preliminary version of parts of the 1975 book.

Now flash backwards some twenty years, to 1952. Feyerabend, in Vienna, arranged to go to Cambridge to study under Wittgenstein. Wittgenstein died in that year. So Feyerabend went to work with Popper at the LSE. His autobiography, *Killing Time*, tells how he quickly established a reputation in the English-speaking world and was offered a job at Berkeley, where he took up a position in the autumn of 1958. He tells how America opened up his life in a way that England never could.

Kuhn and Feyerabend

In his autobiography Feyerabend never even mentions that he soon began having intense conversations with Thomas Kuhn, in which, together, they hatched the idea of incommensurability that was soon to take the world by storm.[15] He rightly connects his thoughts on incommensurability with N. R. Hanson's wonderful but now largely forgotten book, *Patterns of Discovery*, a book that insisted that observational statements are theory-loaded, so that a change in theory implies a change in the meaning even of reports of observations.[16] But he did read a draft of Kuhn's *Structure*

---

10. Ibid. p. 129.

11. Ibid. p. 125.

12. *Killing Time*, p. 128.

13. 'Consolations for the Specialist', in *Criticism and the Growth of Knowledge*, pp. 197–230, on p. 229.

14. Paul Feyerabend, 'Against Method: Outline of an Anarchistic Theory of Knowledge', *Minnesota Studies in the Philosophy of Science* 4 (1970): pp. 17–130.

15. However the pro/anti Kuhn paper, 'Consolations' (see note 13), begins on p. 197 with the statement: 'In the years 1960 and 1961, when Kuhn was a member of the philosophy department at the University of California in Berkeley, I had the good fortune of being able to discuss with him various aspects of science. I have profited enormously from these discussions and I have looked at science in a new way ever since.'

16. N. R. Hanson, *Patterns of Discovery*, Cambridge, 1958.

'around 1960'.[17] Kuhn as a person is not mentioned until, at the end of the autobiography, he becomes 'my old friend'.[18] In the autobiography there is a photo of the two men sitting side by side, in a café near Zurich.

In other work, of course, he grants that Kuhn had the most important ideas about science outside the LSE circle. (Hanson, who did stunt aerobatics among many other things, died when he crashed his plane in 1967 at the age of forty-two.) By the preface to this third edition of *AM*, you will finally find him speaking of 'Kuhn's masterpiece'. In the autobiography, conversely, Feyerabend has many kind things to say about men such as Rudolf Carnap and Herbert Feigl, one-time logical positivists from Vienna, who together with other German-speaking immigrants and refugees changed the face of American philosophy forever. Elsewhere, he was really quite rude about the Vienna Circle. How about 'rodents of neopositivism' in the 1988 Preface to the Chinese edition?[19]

The differences between the personalities of Kuhn and Feyerabend were profound. One of them is best indicated by two adjectives which out of context would be condescending. Kuhn was dogged, and Feyerabend was flighty. Kuhn gnawed at incommensurability for the rest of his life, and left as yet unpublished material which, in my personal judgement, goes wrong in an attempt to produce a theory of incommensurability that would suit linguists and cognitive scientists. Feyerabend revelled in his off-the-wall illustration of the incommensurability of archaic and ancient Greek systems of thought.

Anything goes

Feyerabend will be forever cursed by a statement of his own making, and for which he is fully responsible, the notorious aphorism 'anything goes'. In the Chinese preface, he says it is 'the terrified exclamation of a rationalist who takes a closer look at history'. Yet he would sometimes argue, not in terror but with delight, that even Lakatos's methodology shared with Feyerabend's anti-methodology 'a position of "anything goes"'.[20]

---

17. *Killing Time*, p. 141.

18. Ibid. p.162.

19. In case you think he could not possibly have meant the Vienna Circle by this term of abuse, compare the passage here with the diagram in a letter to Lakatos, *For and Against Method*, p. 245.

20. Ibid., p. 229.

Since the aphorism is often taken to be anti-science, a sort of New Age waffle, we must emphasize that Feyerabend never meant for one minute that anything *except* the scientific method (whatever that is) 'goes'. He meant that lots of ways of getting on, *including* the innumerable methods of the diverse sciences, 'go'. He also meant that an anti-rationalist, like himself, was perfectly entitled to use rationalist arguments to discomfit the rationalists whom he opposed. What he did dislike was any kind of intellectual or ideological hegemony. His favoured text was Mill's *On Liberty*, even if his preferred style was Dada. Single-mindedness in pursuit of any goal, including truth and understanding, yields great rewards. But single vision is folly if it makes you think you see (or even glimpse) *the* truth, the one and only truth. Hence the need for the counter-irritant maxim 'anything goes'.

Anarchism and Dada

For some time Feyerabend cheerfully accepted Lakatos's label 'anarchist'. On 10 October 1970 he wrote to Lakatos that he considered saying in the Preface to *Against Method*: 'I am for anarchism in *thinking*, in one's *private life*, BUT NOT in *public life*.'[21] He did not insert that thought, but he went one better: in 1975, *Against Method* had a subtitle, *Outlines of an anarchistic* theory of knowledge*. Yes, with a footnote to the subtitle, directing the reader to explanations in the text of the term 'anarchism', including the passage that I shall quote in a moment.

The subtitle was abandoned in 1988, although the initial discussion of anarchism in the Introduction is much the same. Moreover, that footnote to the subtitle referred to a chapter that was deleted, though to some extent pasted back here and there in 1988. An important part of that chapter is a discussion of the relation between Dada and intellectual anarchism. The footnote to the subtitle also referred to a very long footnote 12 in the Introduction to the 1975 book. Feyerabend wrote it out in a letter to Lakatos, 7 August 1972.[22] Since it was deleted in 1988, I shall quote it here.

When choosing the term 'anarchism' for my enterprise I simply followed general usage. However anarchism, as it has been practised in the past and as it is being practised today by an ever increasing number of people has

---

21. Ibid., p. 219.
22. Ibid., pp. 294–5.

features I am not prepared to support. It cares little for human lives and human happiness (except for the lives and the happiness of those who belong to some special group); and it contains precisely the kind of Puritanical dedication and seriousness which I detest. (There are some exquisite exceptions such as Cohn-Bendit,[23] but they are in the minority.) It is for these reasons that I now prefer to use the term *Dadaism*. A Dadaist would not hurt a fly – let alone a human being. A Dadaist is utterly unimpressed by any serious enterprise and he smells a rat whenever people stop smiling and assume that attitude and those facial expressions which indicate that something important is about to be said. A Dadaist is convinced that a worthwhile life will arise only when we start taking things *lightly* and when we remove from our speech the profound but already putrid meanings it has accumulated over the centuries ('search for truth'; 'defence of justice'; 'passionate concern'; etc., etc.). A Dadaist is prepared to initiate joyful experiments even in those domains where change and experimentation seem to be out of the question (example: the basic functions of language). I hope that having read the pamphlet[24] the reader will remember me as a flippant Dadaist and *not* as a serious anarchist.

Let us remember him, then, as a Dadaist. Lakatos objected to the claim that a Dadaist would never hurt a fly: sometimes a Dadaist has to do harm when it is the lesser of two evils. Lakatos undoubtedly had in mind a controversial incident in his own past as a resistance fighter, when he compelled his cell to order a young woman to take poison because if caught she would compromise the group. Feyerabend said he accepted Lakatos's criticism. But the argument seems to me not to harm Feyerabend's footnote but, rather, to show that there are times when Dada is not enough. Feyerabend had the moral luck to have been physically injured in war but

---

23. Daniel Cohn-Bendit (born 1945) was a central personality in the Paris student uprisings, May 1968, denounced alike by the Gaullist right and the communist left. In the 1970s he edited a German magazine of anarchist orientation, which increasingly moved toward environmental politics. In 1994 he was elected to the European Parliament as a green, and in 1999 he became leader of the French Green Party. In the June 2009 elections to the European Parliament he was extraordinarily popular with French voters. The right despises him as favouring immigrants, lessening penalties on drugs, and general welfarism, while the left hates his policies of armed intervention in the former Yugoslavia and his fierce support for the European Union.

24. The 'pamphlet' is of course *Against Method*: this is the word Feyerabend used in writing to Lakatos, and he left the paragraph intact when he put it in to the book.

(so he made out) never morally touched. He had the privilege of being able to practise a kind of Dada throughout his life. But it is important to insist that Dada implies passion, not indifference. It may help to understand this by quoting 'a letter to the reader' which was intended to precede his last book, *The Conquest of Abundance*. Although written for another purpose, it can usefully be read by those embarking on *Against Method*.[25]

FEYERABEND'S LAST LETTER

Dear reader,

In a few pages you will find a story written in a style you may be familiar with. There are facts and generalizations therefrom, there are arguments and there are lots of footnotes. In other words, you will find a (perhaps not very outstanding) example of a scholarly essay. Let me therefore warn you that it is not my intention to inform, or to establish some truth. What I want to do is to change your attitude. I want you to sense chaos where at first you noticed an orderly arrangement of well behaved things and processes. It is clear that only a trick can get me from my starting point – the footnote-heavy essay I just mentioned – to where I would like you, the reader, to arrive.

My trick is to present events which dissolve the circumstances that made them happen. Given the circumstances the events are absurd, unheard of, frightening, evil – they simply do not make sense. I take a closer look at the circumstances and find features that may be regarded as anticipations of the event. The features are not unknown; they are not hidden either; however, they can be read in a variety of ways and only some readings create trouble.

The absurdity is therefore not laid out in advance; it is created by living in a certain way – and so is the sense perceived by those who produce the disruptive event. What is interesting is that both parties use the same material; they start from the same life, but they continue it in different directions. (The same applies to the scholars who years and even centuries later try to figure out 'what really happened'.)

I conclude that the life we lead is ambiguous. It contains not only one future, but many, and it contains them neither ready-made nor as possibilities that can be turned into any direction. It is not at all different

---

25. The letter was found by Grazia Borromini-Feyerabend on a disc on 11 October 1999, and was first printed in the London Review of Books, 22 June 2000, p. 28.

from a movie, or a specially constructed play. Imagine such a play. It has gone on for about forty minutes. You know the characters, you have become accustomed to their idiosyncrasies, and you are already tired of their peculiar habits. Now they stand before you with their familiar gestures and it seems that nothing interesting is ever going to happen – when suddenly, because of a trick used by the writer, the 'reality' you perceived turns out to be a chimaera. (Alfred Hitchcock, Anthony Shaffer and Ira Levin are masters of this kind of switch.) Looking back you can now say that things were not what they seemed to be, and looking forward with the experience in mind you will regard any clear and definite arrangement with suspicion, on the stage, and elsewhere. Also, your suspicion will be the greater the more solid the initial story seemed to be. This is why I have chosen a scholarly essay as my starting point.

It is very important not to let this suspicion deteriorate into a truth, or a theory, for example into a theory with the principle: things are never what they seem to be. Reality, or Being, or God, or whatever it is that sustains us cannot be captured that easily. The problem is not why we are so often confused; the problem is why we seem to possess useful and enlightening knowledge.

You must also resist the temptation to classify what I say by giving it a well-established name, for example the name of relativism. Relativism as defined by philosophers and sociologists is much too definite a view to fit the situation – unless it is regarded as a passing chimaera, or as a rule of thumb. You cannot even deny the existence of eternal truths unless the denial is again meant as a cautionary hint given to those visiting the theatre of life. Is argument without a purpose? No, it is not; it accompanies us on our journey without tying it to a fixed road. Is there a way of identifying what is going on? There are many ways and we are using them all the time though often believing that they are part of a stable framework which encompasses everything. Is there a name for an attitude or a view like this? Yes, if names are that important I can easily provide one – mysticism – though it is a mysticism that uses examples, arguments, tightly reasoned passages of text, scientific theories and experiments to raise itself into consciousness.

This, my dear reader, is the warning I want you to remember from time to time, and especially when the story seems to become so definite that it almost turns into a clearly thought out and precisely structured point of view.

Preface

In 1970 Imre Lakatos, one of the best friends I ever had, cornered me at a party. 'Paul,' he said, 'you have such strange ideas. Why don't you write them down? I shall write a reply, we publish the whole thing and I promise you – we shall have lots of fun.' I liked the suggestion and started working. The manuscript of my part of the book was finished in 1972 and I sent it to London. There it disappeared under rather mysterious circumstances. Imre Lakatos, who loved dramatic gestures, notified Interpol and, indeed, Interpol found my manuscript and returned it to me. I reread it and made some final changes. In February 1974, only a few weeks after I had finished my revision, I was informed of Imre's death. I published my part of our common enterprise without his response. A year later I published a second volume, *Science in a Free Society*, containing additional material and replies to criticism.

This history explains the form of the book. It is not a systematic treatise; it is a letter to a friend and addresses his idiosyncrasies. For example, Imre Lakatos was a rationalist, hence rationalism plays a large role in the book. He also admired Popper and therefore Popper occurs much more frequently than his 'objective importance' would warrant. Imre Lakatos, somewhat jokingly, called me an anarchist, and I had no objection to putting on the anarchist's mask. Finally, Imre Lakatos loved to embarrass serious opponents with jokes and irony and so I, too, occasionally wrote in a rather ironical vein. An example is the end of Chapter 1: 'anything goes' is not a 'principle' I hold – I do not think that 'principles' can be used and fruitfully discussed outside the concrete research situation they are supposed to affect – but the terrified exclamation of a rationalist who takes a closer look at history. Reading the many thorough, serious, longwinded and thoroughly misguided criticisms I received after publication of the first English edition I often recalled my exchanges with Imre; how we would both have laughed had we been able to read these effusions together.

The new edition merges parts of *Against Method* with excerpts from *Science in a Free Society*. I have omitted material no longer of interest, added a chapter on the trial of Galileo and a chapter on the notion of reality that seems to be required by the fact that knowledge is part of a complex historical process, eliminated mistakes, shortened the argument wherever possible and freed it from some of its earlier idiosyncrasies.

Again I want to make two points: first, that science can stand on its own feet and does not need any help from rationalists, secular humanists, Marxists and similar religious movements; and, secondly, that non-scientific cultures, procedures and assumptions can also stand on their own feet and should be allowed to do so, if this is the wish of their representatives. Science must be protected from ideologies; and societies, especially democratic societies, must be protected from science. This does not mean that scientists cannot profit from a philosophical education and that humanity has not and never will profit from the sciences. However, the profits should not be imposed; they should be examined and freely accepted by the parties of the exchange. In a democracy scientific institutions, research programmes, and suggestions must therefore be subjected to public control, there must be a separation of state and science just as there is a separation between state and religious institutions, and science should be taught as one view among many and not as the one and only road to truth and reality. There is nothing in the nature of science that excludes such institutional arrangements or shows that they are liable to lead to disaster.

None of the ideas that underlie my argument is new. My interpretation of scientific knowledge, for example, was a triviality for physicists like Mach, Boltzmann, Einstein and Bohr. But the ideas of these great thinkers were distorted beyond recognition by the rodents of neopositivism and the competing rodents of the church of 'critical' rationalism. Lakatos was, after Kuhn, one of the few thinkers who noticed the discrepancy and tried to eliminate it by means of a complex and very interesting theory of rationality. I don't think he has succeeded in this. But the attempt was worth the effort; it has led to interesting results in the history of science and to new insights into the limits of reason. I therefore dedicate also this second, already much more lonely version of our common work to his memory.

Earlier material relating to the problems in this book is now collected in my *Philosophical Papers*.[1] *Farewell to Reason*[2] contains historical material, especially from the early history of rationalism in the West and applications to the problems of today.

Berkeley, September 1987

---

1. 2 vols, Cambridge, 1981.
2. London, 1987.

Introduction to the Chinese Edition

This book proposes a thesis and draws consequences from it. The thesis is: *the events, procedures and results that constitute the sciences have no common structure*; there are no elements that occur in every scientific investigation but are missing elsewhere. Concrete developments (such as the overthrow of steady state cosmologies and the discovery of the structure of DNA) have distinct features and we can often explain why and how these features led to success. But not every discovery can be accounted for in the same manner, and procedures that paid off in the past may create havoc when imposed on the future. Successful research does not obey general standards; it relies now on one trick, now on another; the moves that advance it and the standards that define what counts as an advance are not always known to the movers. Far-reaching changes of outlook, such as the so-called 'Copernican Revolution' or the 'Darwinian Revolution', affect different areas of research in different ways and receive different impulses from them. A theory of science that devises standards and structural elements for *all* scientific activities and authorizes them by reference to 'Reason' or 'Rationality' may impress outsiders – but it is much too crude an instrument for the people on the spot, that is, for scientists facing some concrete research problem.

In this book I try to support the thesis by historical examples. Such support does not *establish* it; it makes it *plausible* and the way in which it is reached indicates how future statements about 'the nature of science' may be undermined: given any rule, or any general statement about the sciences, there always exist developments which are praised by those who support the rule but which show that the rule does more damage than good.

One consequence of the thesis is that *scientific successes cannot be explained in a simple way*. We cannot say: 'the structure of the atomic nucleus was found because people did A, B, C . . .' where A, B and C are procedures which can be understood independently of their use in nuclear physics. All we can do is to give a historical account of the details, including social circumstances, accidents and personal idiosyncrasies.

Another consequence is that *the success of 'science' cannot be used as an argument for treating as yet unsolved problems in a standardized way*. That could be done only if there are procedures that can be detached from particular research situations and whose presence guarantees success.

The thesis says that there are no such procedures. Referring to the success of 'science' in order to justify, say, quantifying human behaviour is therefore an argument without substance. Quantification works in some cases, fails in others; for example, it ran into difficulties in one of the apparently most quantitative of all sciences, celestial mechanics (special region: stability of the planetary system) and was replaced by qualitative (topological) considerations.

It also follows that *'non-scientific' procedures cannot be pushed aside by argument.* To say: 'the procedure you used is non-scientific, therefore we cannot trust your results and cannot give you money for research' assumes that 'science' is successful and that it is successful because it uses uniform procedures. The first part of the assertion ('science is always successful') is not true, if by 'science' we mean things done by scientists – there are lots of failures also. The second part – that successes are due to uniform procedures – is not true because there are no such procedures. Scientists are like architects who build buildings of different sizes and different shapes and who can be judged only *after* the event, i.e. only after they have finished their structure. It may stand up, it may fall down – nobody knows.

But if scientific achievements can be judged only after the event and if there is no abstract way of ensuring success beforehand, then there exists no special way of weighing scientific promises either – scientists are no better off than anybody else in these matters, they only know more details. This means that *the public can participate in the discussion without disturbing existing roads to success* (there are no such roads). In cases where the scientists' work affects the public it even *should* participate: first, because it is a concerned party (many scientific decisions affect public life); secondly, because such participation is the best scientific education the public can get – a full democratization of science (which includes the protection of minorities such as scientists) is not in conflict with science. It is in conflict with a philosophy, often called 'Rationalism', that uses a frozen image of science to terrorize people unfamiliar with its practice.

A consequence to which I allude in Chapter 19 and which is closely connected with its basic thesis is that *there can be many different kinds of science.* People starting from different social backgrounds will approach the world in different ways and learn different things about it. People survived millennia before Western science arose; to do this they had to know their surroundings up to and including elements of astronomy. 'Several thousand Cuahuila Indians never exhausted the natural resources of a desert region in South California, in which today only a handful of white families manage to subsist. They lived in a land of plenty, for in this apparently completely barren territory, they were familiar with no less than sixty kinds of edible plants and twenty-eight others of narcotic, stimulant or medical properties'.[1] The knowledge that preserves the lifestyles of nomads was acquired and is preserved in a non-scientific way ('science' now being modern natural science). Chinese technology for a long time lacked any Western-scientific underpinning and yet it was far ahead of contemporary Western technology. It is true that Western science now reigns supreme all over the globe; however, the reason was not insight in its 'inherent rationality' but power play (the colonizing nations imposed their ways of living) and the need for weapons: Western science so far has created the most efficient instruments of death. The remark that without Western science many 'Third World nations' would be starving is correct but one should add that the troubles were created, not alleviated, by earlier forms of 'development'. It is also true that Western medicine helped eradicate parasites and some infectious diseases, but this does not show that Western science is the only tradition that has good things to offer and that other forms of inquiry are without any merit whatsoever. *First-world science is one science among many*; by claiming to be more it ceases to be an instrument of research and turns into a (political) pressure group. More on these matters can be found in my book *Farewell to Reason*.[2]

My main motive in writing that book was humanitarian, not intellectual. I wanted to support people, not to 'advance knowledge'. People all over the world have developed ways of surviving in partly dangerous, partly agreeable surroundings. The stories they told and the activities they engaged in enriched their lives, protected them and gave them meaning. The 'progress of knowledge and civilization' – as the process of pushing Western ways and values into all corners of the globe is being called – destroyed these wonderful products of human ingenuity and compassion without a single glance in their direction. 'Progress of knowledge' in many places meant killing of minds. Today old traditions are being revived and

---

1. C. Lévi-Strauss, *The Savage Mind*, London, 1966, pp. 4f.
2. London, 1987.

people try again to adapt their lives to the ideas of their ancestors. I have tried to show, by an analysis of the apparently hardest parts of science, the natural sciences, that science, properly understood, has no argument against such a procedure. There are many scientists who act accordingly. Physicians, anthropologists and environmentalists are starting to adapt their procedures to the values of the people they are supposed to advise. I am not against a science so understood. Such a science is one of the most wonderful inventions of the human mind. But I am against ideologies that use the name of science for cultural murder.

Preface to the Third Edition

Many things have happened since I first published *Against Method* (AM for short). There have been dramatic political, social and ecological changes. Freedom has increased – but it has brought hunger, insecurity, nationalistic tensions, wars and straightforward murder. World leaders have met to deal with the deterioration of our resources; as is their habit, they have made speeches and signed agreements. The agreements are far from satisfactory; some of them are a sham. However, at least verbally, the environment has become a world-wide concern. Physicians, developmental agents, priests working with the poor and disadvantaged have realized that these people know more about their condition than a belief in the universal excellence of science or organized religion had assumed and they have changed their actions and their ideas accordingly (liberation theology; primary environmental care, etc.). Many intellectuals have adapted what they have learned at universities and special schools to make their knowledge more efficient and more humane.

On a more academic level historians (of science, of culture) have started approaching the past in its own terms. Already in 1933, in his inaugural lecture at the Collège de France, Lucien Febvre had ridiculed writers who, 'sitting at their desks, behind mountains of paper, having closed and covered their windows', made profound judgements about the life of landholders, peasants and farmhands. In a narrow field historians of science tried to reconstruct the distant and the more immediate past without distorting it by modern beliefs about truth (fact) and rationality. Philosophers then concluded that the various forms of rationalism that had offered their services had not only produced chimaeras but would have damaged the sciences had they been adopted as guides. Here Kuhn's masterpiece played a decisive role.[1] It led to new ideas. Unfortunately it also encouraged lots of trash. Kuhn's main terms ('paradigm', 'revolution', 'normal science', 'prescience', 'anomaly', 'puzzle-solving', etc.) turned up in various forms of pseudoscience while his general approach confused many writers: finding that science had been freed from the fetters of a dogmatic logic and epistemology they tried to tie it down again, this time with sociological ropes. That trend lasted well into the early seventies. By contrast there are now historians and sociologists who concentrate

---

1. *The Structure of Scientific Revolutions*, Chicago, 1962.

on particulars and allow generalities only to the extent that they are supported by sociohistorical connections. 'Nature', says Bruno Latour, referring to 'science in the making' is 'the consequence of [a] settlement' of 'controversies'.[2] Or, as I wrote in the first edition of AM: 'Creation of a *thing*, and creation plus full understanding of a *correct idea* of the thing, *are very often parts of one and the same indivisible process* and cannot be separated without bringing the process to a stop.'[3]

Examples of the new approach are Andrew Pickering, *Constructing Quarks*, Peter Galison, *How Experiments End*, Martin Rudwick, *The Great Devonian Controversy*, Arthur Fine, *The Shaky Game* and others.[4] There are studies of the various traditions (religious, stylistic, patronage, etc.) that influenced scientists and shaped their research;[5] they show the need for a far more complex account of scientific knowledge than that which had emerged from positivism and similar philosophies. On a more general level we have the older work of Michael Polanyi and then Putnam, van Fraassen, Cartwright, Marcello Pera[6] and, yes, Imre Lakatos, who was sufficiently optimistic to believe that history herself – a lady he took very seriously – offered simple rules of theory evaluation.

In sociology the attention to detail has led to a situation where the problem is no longer why and how 'science' changes but how it keeps together. Philosophers, philosophers of biology especially, suspected for some time that there is not one entity 'science' with clearly defined principles but that science contains a great variety of (high-level theoretical, phenomenological, experimental) approaches and that even a particular science such as physics is but a scattered collection of subjects (elasticity, hydrodynamics, rheology, thermodynamics, etc., etc.) each one containing contrary tendencies (example: Prandtl vs Helmholtz, Kelvin, Lamb, Rayleigh; Truesdell vs Prandtl; Birkhoff vs 'physical commonsense'; Kinsman illustrating all trends – in hydrodynamics). For some authors this is not only a fact; it is also desirable.[7] Here again I contributed, in a small way, in Chapters 3, 4 and 11 of AM,[8] in section 6

---

2. *Science in Action*, Milton Keynes, 1987, pp. 4 and 98f.
3. London, 1975, p. 26, repeated on p. 17 of the present edition – original emphasis.
4. All Chicago University Press.
5. An example is Mario Biagioli, *Galileo Courtier*, forthcoming.
6. *Science and Rhetoric*, forthcoming.
7. J. Dupré, 'The Disunity of Science', *Mind* 92, 1983.
8. Present edition. Taken over unamended from first edition.

of my contribution to Lakatos and Musgrave's *Criticism and the Growth of Knowledge* (criticism of the uniformity of paradigms in Kuhn)[9] and already in 1962, in my contribution to the *Delaware Studies for the Philosophy of Science*.[10]

Unity further disappears when we pay attention not only to breaks on the theoretical level, but to experiment and, especially, to modern laboratory science. As Ian Hacking has shown in his pathbreaking essay *Representing and Intervening*[11] and as emerges from Pickering's *Science as Practice and Culture*,[12] terms such as 'experiment' and 'observation' cover complex processes containing many strands. 'Facts' come from negotiations between different parties and the final product – the published report – is influenced by physical events, dataprocessors, compromises, exhaustion, lack of money, national pride and so on. Some microstudies of laboratory science resemble the 'New Journalism' of Jimmy Breslin, Guy Talese, Tom Wolfe and others; researchers no longer sit back and read the papers in a certain field; they are not content with silent visits to laboratories either – they walk right in, engage scientists in conversation and make things happen (Kuhn and his collaborators started the procedure in their interviews for the history of quantum mechanics). At any rate – we are a long way from the old (Platonic) idea of science as a system of statements growing with experiment and observation and kept in order by lasting rational standards.

AM is still partly proposition oriented; however, I also had my sane moments. My discussion of incommensurability, for example, does not 'reduce the difference to one of theory' as Pickering writes.[13] It includes art forms, perceptions (a large part of Chapter 16 is about the transition from Greek geometric art and poetry to the classical period), and stages of child development and asserts 'that the views of scientists and especially their views on basic matters are often as different from each other as are the ideologies of different cultures'.[14] In this connection I examined the practical aspects of logic, the way, that is, in which ideas are related to

---

9. I. Lakaros and A. Musgrave (eds), *Criticism and the Growth of Knowledge*, Cambridge, 1965.

10. 'How to be a Good Empiricist', *Delaware Studies*, Vol. 2, 1963.

11. Cambridge, 1983.

12. A. Pickering (ed.), *Science as Practice and Culture*, Chicago, 1992.

13. ibid., p. 10.

14. AM, first edition, p. 274.

each other in ongoing research rather than in the finished products (if there ever are such products). My discussion of the many events that constitute what is being observed[15] and especially my discussion of Galileo's telescopic discoveries[16] agree with the requirements of the new laboratory sociology except that Galileo's 'laboratory' was rather small by comparison. This case shows, incidentally, that like the older philosophies of science the new microsociology is not a universal account but a description of prominent aspects of a special period. It does not matter. A universal description of science at any rate can at most offer a list of events.[17] It was different in antiquity.

It is clear that the new situation requires a new philosophy and, above all, new terms. Yet some of the foremost researchers in the area are still asking themselves whether a particular piece of research produces a 'discovery', or an 'invention', or to what extent a (temporary) result is 'objective'. The problem arose in quantum mechanics; it is also a problem for classical science. Shall we continue using outmoded terms to describe novel insight or would it not be better to start using a new language? And wouldn't poets and journalists be of great help in finding such a language?

Secondly, the new situation again raises the question of 'science' vs democracy. For me this was the most important question. 'My main reason for writing the book', I say in the Introduction to the Chinese Edition,[18] 'was humanitarian, not intellectual. I wanted to support people, not to "advance knowledge".' Now if science is no longer a unit, if different parts of it proceed in radically different ways and if connections between these ways are tied to particular research episodes, then scientific projects have to be taken individually. This is what government agencies started doing some time ago. In the late sixties 'the idea of a comprehensive science policy was gradually abandoned. It was realized that science was not one but many enterprises and that there could be no single policy for the support of all of them.'[19] Government agencies no longer finance 'science', they finance particular projects. But then the word 'scientific'

---

15. Ibid., pp. 149ff. Reprinted in the present edition.

16. Chapters 8 to 10 of the present edition.

17. Cf. my contribution to the 1992 Erasmus Symposium, 'Has the Scientific View of the World a Special Status Compared With Other Views?', forthcoming.

18. Contained in the present edition.

19. J. Ben-David, *Scientific Growth*, Berkeley, 1991, p. 525.

can no longer exclude 'unscientific' projects – we have to look at matters in detail. Are the new philosophers and sociologists prepared to consider this consequence of their research?

There have been many other changes. Medical researchers and technologists have not only invented useful instruments (such as those employing the principles of fibre optics which in many contexts replace the more dangerous methods of X-ray diagnostics) but have become more open towards new (or older) ideas. Only twenty years ago the idea that the mind affects physical well-being, though supported by experience, was rather unpopular – today it is mainstream. Malpractice suits have made physicians more careful, sometimes too careful for the good of their patients, but they have also forced them to consult alternative opinions. (In Switzerland a belligerent plurality of views is almost part of culture – and I used it when arranging public confrontations between hardheaded scientists and 'alternative' thinkers.[20]) However, here as elsewhere, simple philosophies, whether of a dogmatic or a more liberal kind, have their limits. *There are no general solutions.* An increased liberalism in the definition of 'fact' can have grave repercussions,[21] while the idea that truth is concealed and even perverted by the processes that are meant to establish it makes excellent sense.[22] I therefore again warn the reader that I don't have the intention of replacing 'old and dogmatic' principles by 'new and more libertarian ones'. For example, I am neither a populist for whom an appeal to 'the people' is the basis of all knowledge, nor a relativist for whom there are no 'truths as such' but only truths for this or that group and/or individual. All I say is that non-experts often know more than experts *and should therefore be consulted*, and that prophets of truth (including those who use arguments) more often than not are carried along by a vision that clashes with the very events the vision is supposed to be exploring. There exists ample evidence for both parts of this assertion.

A case I already mentioned is development: professionals dealing with the ecological, social and medical parts of developmental aid have by

---

20. Cf. the series edited by Christian Thomas and myself and published by the Verlag der Fachvereine, Zurich, 1983–87.

21. Cf. Peter W. Huber, *Galileo's Revenge*, New York, 1991.

22. For a fictional account, cf. Tom Wolfe's *The Bonfire of the Vanities*, New York, 1987.

now realized that the imposition of 'rational' or 'scientific' procedures, though occasionally beneficial (removal of some parasites and infectious diseases), can lead to serious material and spiritual problems. They did not abandon what they had learned in their universities, however; they combined this knowledge with local beliefs and customs and thereby established a much needed link with the problems of life that surround us everywhere, in the First, Second, and Third Worlds.

The present edition contains major changes (Chapter 19 and part of Chapter 16 have been rewritten, the old Chapter 20 has been omitted), additions (a paragraph here, a paragraph there), stylistic changes (I hope they are improvements) and corrections as well as additions in the references. As far as I am concerned the main ideas of the essay (i.e. the ideas expressed in italics in the Introduction to the Chinese Edition) are rather trivial and appear trivial when expressed in suitable terms. I prefer more paradoxical formulations, however, for nothing dulls the mind as thoroughly as hearing familiar words and slogans. It is one of the merits of deconstruction to have undermined philosophical commonplaces and thus to have made some people think. Unfortunately it affected only a small circle of insiders and it affected them in ways that are not always clear, not even to them. That's why I prefer Nestroy, who was a great, popular and funny deconstructeur, while Derrida, for all his good intentions, can't even tell a story.

Rome, July 1992

Analytical Index

Being a Sketch of the Main Argument

5   **33**

No theory ever agrees with all the facts in its domain, yet it is not always the theory that is to blame. Facts are constituted by older ideologies, and a clash between facts and theories may be proof of progress. It is also a first step in our attempt to find the principles implicit in familiar observational notions.

6   **49**

As an example of such an attempt I examine the tower argument *which the Aristotelians used to refute the motion of the earth. The argument involves* natural interpretations – *ideas so closely connected with observations that it needs a special effort to realize their existence and to determine their content. Galileo identifies the natural interpretations which are inconsistent with Copernicus and replaces them by others.*

7   **61**

The new natural interpretations constitute a new and highly abstract observation language. They are introduced and concealed *so that one fails to notice the change that has taken place (method of anamnesis). They contain the idea of the* relativity of all motion *and the* law of circular inertia.

8   **74**

In addition to natural interpretations, Galileo also changes sensations *that seem to endanger Copernicus. He admits that there are such sensations, he praises Copernicus for having disregarded them, he claims to have removed them with the help of the* telescope. *However, he offers no theoretical reasons why the telescope should be expected to give a true picture of the sky.*

9   **83**

Nor does the initial experience *with the telescope provide such reasons. The first telescopic observations of the sky are indistinct, indeterminate, contradictory and in conflict with what everyone can see with his unaided eyes. And the only theory that could have helped to separate telescopic illusions from veridical phenomena was refuted by simple tests.*

Introduction

Science is an essentially anarchic enterprise: theoretical anarchism is more humanitarian and more likely to encourage progress than its law-and-order alternatives.

Ordnung ist heutzutage meistens dort, wo nichts ist. Es ist eine Mangelerscheinung.

Brecht

The following essay is written in the conviction that *anarchism*, while perhaps not the most attractive *political* philosophy, is certainly excellent medicine for *epistemology*, and for the *philosophy of science*.

The reason is not difficult to find.

'History generally, and the history of revolution in particular, is always richer in content, more varied, more many-sided, more lively and subtle than even' the best historian and the best methodologist can imagine.[1] History is full of 'accidents and conjunctures and curious juxtapositions of events'[2] and it demonstrates to us the 'complexity of human change and the unpredictable character of the ultimate consequences of any given act or decision of men'.[3] Are we really to believe that the naive and simple-minded rules which methodologists take as their guide are capable of accounting for such a 'maze of interactions'?[4] And is it not

---

1. 'History as a whole, and the history of revolutions in particular, is always richer in content, more varied, more multiform, more lively and ingenious than is imagined by even the best parties, the most conscious vanguards of the most advanced classes' (V.I. Lenin, 'Left-Wing Communism – An Infantile Disorder', *Selected Works*, Vol. 3, London, 1967, p. 401). Lenin is addressing parties and revolutionary vanguards rather than scientists and methodologists; the lesson, however, is the same. Cf. footnote 5.

2. Herbert Butterfield, *The Whig Interpretation of History*, New York, 1965, p. 66.

3. Ibid., p. 21.

4. Ibid., p. 25, cf. Hegel, *Philosophie der Geschichte, Werke*, Vol. 9, ed. Edward Gans, Berlin, 1837, p. 9: 'But what experience and history teach us is this, that nations and governments have never learned anything from history, or acted according to rules that might have derived from it. Every period has such peculiar circumstances, is in such an individual state, that decisions will have to be made, and decisions *can* only be made, in it and out of it.' – 'Very clever'; 'shrewd and very clever'; 'NB' writes Lenin in his marginal notes to this passage. (*Collected Works*, Vol. 38, London, 1961, p. 307.)

clear that successful *participation* in a process of this kind is possible only for a ruthless opportunist who is not tied to any particular philosophy and who adopts whatever procedure seems to fit the occasion?

This is indeed the conclusion that has been drawn by intelligent and thoughtful observers. 'Two very important practical conclusions follow from this [character of the historical process],' writes Lenin,[5] continuing the passage from which I have just quoted. 'First, that in order to fulfil its task, the revolutionary class [i.e. the class of those who want to change either a part of society such as science, or society as a whole] must be able to master *all* forms or aspects of social activity without exception [it must be able to understand, and to apply, not only one particular methodology, but any methodology, and any variation thereof it can imagine] . . .; second [it] must be ready to pass from one to another in the quickest and most unexpected manner.' 'The external conditions', writes Einstein,[6] 'which are set for [the scientist] by the facts of experience do not permit him to let himself be too much restricted, in the construction of his conceptual world, by the adherence to an epistemological system. He, therefore, must appear to the systematic epistemologist as a type of unscrupulous opportunist. . . .' A complex medium containing surprising and unforeseen developments demands complex procedures and defies analysis on the basis of rules which

---

5. Ibid. We see here very clearly how a few substitutions can turn a political lesson into a lesson for *methodology*. This is not at all surprising. Methodology and politics are both means for moving from one historical stage to another. We also see how an individual, such as Lenin, who is not intimidated by traditional boundaries and whose thought is not tied to the ideology of a particular profession, can give useful advice to everyone, philosophers of science included. In the 19th century the idea of an elastic and historically informed methodology was a matter of course. Thus Ernst Mach wrote in his book *Erkenntnis und Irrtum*, Neudruck, Wissenschaftliche Buchgesell-schaft, Darmstadt, 1980, p. 200: 'It is often said that research cannot be taught. That is quite correct, in a certain sense. The schemata of *formal* logic and of *inductive* logic are of little use for the intellectual situations are never exactly the same. But the examples of great scientists are very suggestive.' They are not suggestive because we can abstract rules from them and subject future research to their jurisdiction; they are suggestive because they make the mind nimble and capable of inventing entirely new research traditions. For a more detailed account of Mach's philosophy see my essay *Farewell to Reason*, London, 1987, Chapter 7, as well as Vol. 2, Chapters 5 and 6 of my *Philosophical Papers*, Cambridge, 1981.

6. Albert Einstein, *Albert Einstein: Philosopher Scientist*, ed. P.A. Schilpp, New York, 1951, pp. 683f.

have been set up in advance and without regard to the ever-changing conditions of history.

Now it is, of course, possible to simplify the medium in which a scientist works by simplifying its main actors. The history of science, after all, does not just consist of facts and conclusions drawn from facts. It also contains ideas, interpretations of facts, problems created by conflicting interpretations, mistakes, and so on. On closer analysis we even find that science knows no 'bare facts' at all but that the 'facts' that enter our knowledge are already viewed in a certain way and are, therefore, essentially ideational. This being the case, the history of science will be as complex, chaotic, full of mistakes, and entertaining as the ideas it contains, and these ideas in turn will be as complex, chaotic, full of mistakes, and entertaining as are the minds of those who invented them. Conversely, a little brainwashing will go a long way in making the history of science duller, simpler, more uniform, more 'objective' and more easily accessible to treatment by strict and unchangeable rules.

Scientific education as we know it today has precisely this aim. It simplifies 'science' by simplifying its participants: first, a domain of research is defined. The domain is separated from the rest of history (physics, for example, is separated from metaphysics and from theology) and given a 'logic' of its own. A thorough training in such a 'logic' then conditions those working in the domain; it makes *their actions* more uniform and it freezes large parts of the *historical process* as well. Stable 'facts' arise and persevere despite the vicissitudes of history. An essential part of the training that makes such facts appear consists in the attempt to inhibit intuitions that might lead to a blurring of boundaries. A person's religion, for example, or his metaphysics, or his sense of humour (his *natural* sense of humour and not the inbred and always rather nasty kind of jocularity one finds in specialized professions) must not have the slightest connection with his scientific activity. His imagination is restrained, and even his language ceases to be his own. This is again reflected in the nature of scientific 'facts' which are experienced as being independent of opinion, belief, and cultural background.

It is thus *possible* to create a tradition that is held together by strict rules, and that is also successful to some extent. But is it *desirable* to support such a tradition to the exclusion of everything else? Should we transfer to it the sole rights for dealing in knowledge, so that any result that has been obtained by other methods is at once ruled out of court?

And did scientists ever remain within the boundaries of the traditions they defined in this narrow way? These are the questions I intend to ask in the present essay. And to these questions my answer will be a firm and resounding NO.

There are two reasons why such an answer seems to be appropriate. The first reason is that the world which we want to explore is a largely unknown entity. We must, therefore, keep our options open and we must not restrict ourselves in advance. Epistemological prescriptions may look splendid when compared with other epistemological prescriptions, or with general principles – but who can guarantee that they are the best way to discover, not just a few isolated 'facts', but also some deep-lying secrets of nature? The second reason is that a scientific education as described above (and as practised in our schools) cannot be reconciled with a humanitarian attitude. It is in conflict 'with the cultivation of individuality which alone produces, or can produce, well-developed human beings';[7] it 'maims by compression, like a Chinese lady's foot, every part of human nature which stands out prominently, and tends to make a person markedly different in outline'[8] from the ideals of rationality that happen to be fashionable in science, or in the philosophy of science. The attempt to increase liberty, to lead a full and rewarding life, and the corresponding attempt to discover the secrets of nature and of man, entails, therefore, the rejection of all universal standards and of all rigid traditions. (Naturally, it also entails the rejection of a large part of contemporary science.)

It is surprising to see how rarely the stultifying effect of 'the Laws of Reason' or of scientific practice is examined by professional anarchists. Professional anarchists oppose any kind of restriction and they demand that the individual be permitted to develop freely, unhampered by laws, duties or obligations. And yet they swallow without protest all the severe standards which scientists and logicians impose upon research and upon any kind of knowledge-creating and knowledge-changing activity. Occasionally, the laws of scientific method, or what are thought to be the laws of scientific method by a particular writer, are even integrated into anarchism itself. 'Anarchism is a world concept based upon a mechanical

---

7. John Stuart Mill, 'On Liberty', in *The Philosophy of John Stuart Mill*, ed. Marshall Cohen, New York, 1961, p. 258.

8. Ibid., p. 265.

explanation of all phenomena,' writes Kropotkin.[9] 'Its method of investigation is that of the exact natural sciences ... the method of induction and deduction.' 'It is not so clear,' writes a modern 'radical' professor at Columbia,[10] 'that scientific research demands an absolute freedom of speech and debate. Rather the evidence suggests that certain kinds of unfreedom place no obstacle in the way of science....'

There are certainly some people to whom this is 'not so clear'. Let us, therefore, start with our outline of an anarchistic methodology and a corresponding anarchistic science. There is no need to fear that the diminished concern for law and order in science and society that characterizes an anarchism of this kind will lead to chaos. The human nervous system is too well organized for that.[11] There may, of course, come a time when it will be necessary to give reason a temporary advantage and when it will be wise to defend its rules to the exclusion of everything else. I do not think that we are living in such a time today.[12]

---

9. Peter Alexeivich Kropotkin, 'Modern Science and Anarchism', *Kropotkin's Revolutionary Pamphlets*, ed. R.W. Baldwin, New York, 1970, pp. 150–2. 'It is one of Ibsen's great distinctions that nothing was valid for him but science.' B. Shaw, *Back to Methuselah*, New York, 1921, p. xcvii. Commenting on these and similar phenomena Strindberg writes (*Antibarbarus*): 'A generation that had the courage to get rid of God, to crush the state and church, and to overthrow society and morality, still bowed before Science. And in Science, where freedom ought to reign, the order of the day was "believe in the authorities or off with your head".'

10. R.P. Wolff, *The Poverty of Liberalism*, Boston, 1968, p. 15. For a criticism of Wolff see footnote 52 of my essay 'Against Method', in *Minnesota Studies in the Philosophy of Science*, Vol. 4, Minneapolis, 1970.

11. Even in undetermined and ambiguous situations, uniformity of action is soon achieved and adhered to tenaciously. See Muzafer Sherif, *The Psychology of Social Norms*, New York, 1964.

12. This was my opinion in 1970 when I wrote the first version of this essay. Times have changed. Considering some tendencies in US education ('politically correct' academic menus, etc.), in philosophy (postmodernism) and in the world at large I think that reason should now be given greater weight not because it is and always was fundamental but because it seems to be needed, in circumstances that occur rather frequently today (but may disappear tomorrow), to create a more humane approach.

1

This is shown both by an examination of historical episodes and by an abstract analysis of the relation between idea and action. The only principle that does not inhibit progress is: anything goes.

The idea of a method that contains firm, unchanging, and absolutely binding principles for conducting the business of science meets considerable difficulty when confronted with the results of historical research. We find, then, that there is not a single rule, however plausible, and however firmly grounded in epistemology, that is not violated at some time or other. It becomes evident that such violations are not accidental events, they are not results of insufficient knowledge or of inattention which might have been avoided. On the contrary, we see that they are necessary for progress. Indeed, one of the most striking features of recent discussions in the history and philosophy of science is the realization that events and developments, such as the invention of atomism in antiquity, the Copernican Revolution, the rise of modern atomism (kinetic theory; dispersion theory; stereochemistry; quantum theory), the gradual emergence of the wave theory of light, occurred only because some thinkers either *decided* not to be bound by certain 'obvious' methodological rules, or because they *unwittingly broke* them.

This liberal practice, I repeat, is not just a *fact* of the history of science. It is both reasonable and *absolutely necessary* for the growth of knowledge. More specifically, one can show the following: given any rule, however 'fundamental' or 'rational', there are always circumstances when it is advisable not only to ignore the rule, but to adopt its opposite. For example, there are circumstances when it is advisable to introduce, elaborate, and defend *ad hoc* hypotheses, or hypotheses which contradict well-established and generally accepted experimental results, or hypotheses whose content is smaller than the content of the existing and empirically adequate alternative, or self-inconsistent hypotheses, and so on.[1]

---

1. One of the few thinkers to understand this feature of the development of knowledge was Niels Bohr: '. . . he would never try to outline any finished picture, but would patiently go through all the phases of the development of a problem, starting from some apparent paradox, and gradually leading to its elucidation. In fact, he never

There are even circumstances – and they occur rather frequently – when *argument* loses its forward-looking aspect and becomes a hindrance to progress. Nobody would claim that the teaching of *small children* is exclusively a matter of argument (though argument may enter into it, and should enter into it to a larger extent than is customary), and almost everyone now agrees that what looks like a result of reason – the mastery of a language, the existence of a richly articulated perceptual world, logical ability – is due partly to indoctrination and partly to a process of *growth* that proceeds with the force of natural law. And where arguments do seem to have an effect, this is more often due to their *physical repetition* than to their *semantic content*.

Having admitted this much, we must also concede the possibility of non-argumentative growth in the *adult* as well as in (the theoretical parts of) *institutions* such as science, religion, prostitution, and so on. We certainly cannot take it for granted that what is possible for a small child – to acquire new modes of behaviour on the slightest provocation, to

---

regarded achieved results in any other light than as starting points for further exploration. In speculating about the prospects of some line of investigation, he would dismiss the usual consideration of simplicity, elegance or even consistency with the remark that such qualities can only be properly judged *after* [my italics] the event. . . .' L. Rosenfeld in *Niels Bohr. His Life and Work as seen by his Friends and Colleagues*, S. Rosental (ed.), New York, 1967, p. 117. Now science is never a completed process, therefore it is always 'before' the event. Hence simplicity, elegance or consistency are *never* necessary conditions of (scientific) practice.

Considerations such as these are usually criticized by the childish remark that a contradiction 'entails' everything. But contradictions do not 'entail' anything unless people use them in certain ways. And people will use them as entailing everything only if they accept some rather simple-minded rules of derivation. Scientists proposing theories with logical faults and obtaining interesting results with their help (for example: the results of early forms of the calculus; of a geometry where lines consist of points, planes of lines and volumes of planes; the predictions of the older quantum theory and of early forms of the quantum theory of radiation – and so on) evidently proceed according to different rules. The criticism therefore falls back on its authors unless it can be shown that a logically decontaminated science has better results. Such a demonstration is impossible. Logically perfect versions (if such versions exist) usually arrive only long after the imperfect versions have enriched science by their contributions. For example, wave mechanics was not a 'logical reconstruction' of preceding theories; it was an attempt to preserve their achievements and to solve the physical problems that had arisen from their use. Both the achievements and the problems were produced in a way very different from the ways of those who want to subject everything to the tyranny of 'logic'.

slide into them without any noticeable effort – is beyond the reach of his elders. One should rather expect that catastrophic changes in the physical environment, wars, the breakdown of encompassing systems of morality, political revolutions, will transform adult reaction patterns as well, including important patterns of argumentation. Such a transformation may again be an entirely natural process and the only function of a rational argument may lie in the fact that it increases the mental tension that preceded *and caused* the behavioural outburst.

Now, if there are events, not necessarily arguments, which *cause* us to adopt new standards, including new and more complex forms of argumentation, is it then not up to the defenders of the *status quo* to provide, not just counter-arguments, but also contrary *causes*? ('Virtue without terror is ineffective,' says Robespierre.) And if the old forms of argumentation turn out to be too weak a cause, must not these defenders either give up or resort to stronger and more 'irrational' means? (It is very difficult, and perhaps entirely impossible, to combat the effects of brainwashing by argument.) Even the most puritanical rationalist will then be forced to stop reasoning and to use *propaganda* and *coercion*, not because some of his *reasons* have ceased to be valid, but because the *psychological conditions* which make them effective, and capable of influencing others, have disappeared. And what is the use of an argument that leaves people unmoved?

Of course, the problem never arises quite in this form. The teaching of standards and their defence never consists merely in putting them before the mind of the student and making them as *clear* as possible. The standards are supposed to have maximal *causal efficacy* as well. This makes it very difficult indeed to distinguish between the *logical force* and the *material effect* of an argument. Just as a well-trained pet will obey his master no matter how great the confusion in which he finds himself, and no matter how urgent the need to adopt new patterns of behaviour, so in the very same way a well-trained rationalist will obey the mental image of *his* master, he will conform to the standards of argumentation he has learned, he will adhere to these standards no matter how great the confusion in which he finds himself, and he will be quite incapable of realizing that what he regards as the 'voice of reason' is but a *causal after-effect* of the training he had received. He will be quite unable to discover that the appeal to reason to which he succumbs so readily is nothing but a *political manoeuvre*.

That interests, forces, propaganda and brainwashing techniques play a much greater role than is commonly believed in the growth of our knowledge and in the growth of science, can also be seen from an analysis of the *relation between idea and action*. It is often taken for granted that a clear and distinct understanding of new ideas precedes, and should precede, their formulation and their institutional expression. *First*, we have an idea, or a problem, *then* we act, i.e. either speak, or build, or destroy. Yet this is certainly not the way in which small children develop. They use words, they combine them, they play with them, until they grasp a meaning that has so far been beyond their reach. And the initial playful activity is an essential prerequisite of the final act of understanding. There is no reason why this mechanism should cease to function in the adult. We must expect, for example, that the *idea* of liberty could be made clear only by means of the very same actions, which were supposed to *create* liberty. Creation of a *thing*, and creation plus full understanding of a *correct idea* of the thing, *are very often parts of one and the same indivisible process* and cannot be separated without bringing the process to a stop. The process itself is not guided by a well-defined programme, and cannot be guided by such a programme, for it contains the conditions for the realization of all possible programmes. It is guided rather by a vague urge, by a 'passion' (Kierkegaard). The passion gives rise to specific behaviour which in turn creates the circumstances and the ideas necessary for analysing and explaining the process, for making it 'rational'.

The development of the Copernican point of view from Galileo to the 20th century is a perfect example of the situation I want to describe. We start with a strong belief that runs counter to contemporary reason and contemporary experience. The belief spreads and finds support in other beliefs which are equally unreasonable, if not more so (law of inertia; the telescope). Research now gets deflected in new directions, new kinds of instruments are built, 'evidence' is related to theories in new ways until there arises an ideology that is rich enough to provide independent arguments for any particular part of it and mobile enough to find such arguments whenever they seem to be required. We can say today that Galileo was on the right track, for his persistent pursuit of what once seemed to be a silly cosmology has by now created the material needed to defend it against all those who will accept a view only if it is told in a certain way and who will trust it only if it contains certain magical phrases, called 'observational reports'. And this is not an exception – it is the normal case: theories become clear and 'reasonable' only *after* incoherent parts of them have been used for a long time. Such unreasonable, nonsensical, unmethodical foreplay thus turns out to be an unavoidable precondition of clarity and of empirical success.

Now, when we attempt to describe and to understand developments of this kind in a general way, we are, of course, obliged to appeal to the existing forms of speech which do not take them into account and which must be distorted, misused, beaten into new patterns in order to fit unforeseen situations (without a constant misuse of language there cannot be any discovery, any progress). 'Moreover, since the traditional categories are the gospel of everyday thinking (including ordinary scientific thinking) and of everyday practice, [such an attempt at understanding] in effect presents rules and forms of false thinking and action – false, that is, from the standpoint of (scientific) common sense.'[2] This is how *dialectical thinking* arises as a form of thought that 'dissolves into nothing the detailed determinations of the understanding,'[3] formal logic included.

(Incidentally, it should be pointed out that my frequent use of such words as 'progress', 'advance', 'improvement', etc., does not mean that I claim to possess special knowledge about what is good and what is bad in the sciences and that I want to impose this knowledge upon my readers. *Everyone can read the terms in his own way* and in accordance with the tradition to which he belongs. Thus for an empiricist, 'progress' will mean transition to a theory that provides direct empirical tests for most of its basic assumptions. Some people believe the quantum theory to be a theory of this kind. For others, 'progress' may mean unification and harmony, perhaps even at the expense of empirical adequacy. This is how Einstein viewed the general theory of relativity. *And my thesis is that anarchism helps to achieve progress in any one of the senses one cares to choose.* Even a law-and-order science will succeed only if anarchistic moves are occasionally allowed to take place.)

It is clear, then, that the idea of a fixed method, or of a fixed theory of rationality, rests on too naive a view of man and his social surroundings. To those who look at the rich material provided by history, and who are

---

2. Herbert Marcuse, *Reason and Revolution*, London, 1941, p. 130.
3. Hegel, *Wissenschaft der Logik*, Vol. 1, Hamburg, 1965, p. 6.

not intent on impoverishing it in order to please their lower instincts, their craving for intellectual security in the form of clarity, precision, 'objectivity', 'truth', it will become clear that there is only one principle that can be defended under *all* circumstances and in all stages of human development. It is the principle: *anything goes.*

This abstract principle must now be examined and explained in concrete detail.

2

For example, we may use hypotheses that contradict well-confirmed theories and/or well-established experimental results. We may advance science by proceeding counterinductively.

Examining the principle in concrete detail means tracing the consequences of 'counterrules' which oppose familiar rules of the scientific enterprise. To see how this works, let us consider the rule that it is 'experience', or the 'facts', or 'experimental results' which measure the success of our theories, that agreement between a theory and the 'data' favours the theory (or leaves the situation unchanged) while disagreement endangers it, and perhaps even forces us to eliminate it. This rule is an important part of all theories of confirmation and corroboration. It is the essence of empiricism. The 'counterrule' corresponding to it advises us to introduce and elaborate hypotheses which are inconsistent with well-established theories and/or well-established facts. It advises us to proceed *counterinductively.*

The counterinductive procedure gives rise to the following questions: Is counterinduction more reasonable than induction? Are there circumstances favouring its use? What are the arguments for it? What are the arguments against it? Is perhaps induction always preferable to counterinduction? And so on.

These questions will be answered in two steps. I shall first examine the counterrule that urges us to develop hypotheses inconsistent with accepted and highly confirmed *theories*. Later on I shall examine the counterrule that urges us to develop hypotheses inconsistent with well-established *facts*. The results may be summarized as follows.

In the first case it emerges that the evidence that might refute a theory can often be unearthed only with the help of an incompatible alternative: the advice (which goes back to Newton and which is still very popular today) to use alternatives only when refutations have already discredited the orthodox theory puts the cart before the horse. Also, some of the most important formal properties of a theory are found by contrast, and not by analysis. A scientist who wishes to maximize the empirical content of the views he holds and who wants to understand them as clearly as he possibly can must therefore introduce other views; that is, he must adopt a *pluralistic methodology.* He must compare ideas with other ideas rather than with 'experience' and he must try to improve rather than discard the views that have failed in the competition. Proceeding in this way he will retain the theories of man and cosmos that are found in Genesis, or in the Pimander, he will elaborate them and use them to measure the success of evolution and other 'modern' views. He may then discover that the theory of evolution is not as good as is generally assumed and that it must be supplemented, or entirely replaced, by an improved version of Genesis. Knowledge so conceived is not a series of self-consistent theories that converges towards an ideal view; it is not a gradual approach to the truth. It is rather an ever increasing *ocean of mutually incompatible alternatives*, each single theory, each fairy-tale, each myth that is part of the collection forcing the others into greater articulation and all of them contributing, via this process of competition, to the development of our consciousness. Nothing is ever settled, no view can ever be omitted from a comprehensive account. Plutarch or Diogenes Laertius, and not Dirac or von Neumann, are the models for presenting a knowledge of this kind in which the *history* of a science becomes an inseparable part of the science itself – it is essential for its further *development* as well as for giving *content* to the theories it contains at any particular moment. Experts and laymen, professionals and dilettanti, truth-freaks and liars – they all are invited to participate in the contest and to make their contribution to the enrichment of our culture. The task of the scientist, however, is no longer 'to search for the truth', or 'to praise god', or 'to systematize observations', or 'to improve predictions'. These are but side effects of an activity to which his attention is now mainly directed and which is '*to make the weaker case the stronger*' as the sophists said, *and thereby to sustain the motion of the whole*.

The second 'counterrule' which favours hypotheses inconsistent with *observations, facts and experimental results*, needs no special defence, for there is not a single interesting theory that agrees with all the known facts in its domain. The question is, therefore, not whether counterinductive theories should be *admitted* into science; the question is, rather, whether the *existing* discrepancies between theory and fact should be increased, or diminished, or what else should be done with them.

To answer this question it suffices to remember that observational reports, experimental results, 'factual' statements, either *contain* theoretical assumptions or *assert* them by the manner in which they are used. (For this point cf. the discussion of natural interpretations in Chapters 6ff.) Thus our habit of saying 'the table is brown' when we view it under normal circumstances, with our senses in good order, but 'the table seems to be brown' when either the lighting conditions are poor or when we feel unsure in our capacity of observation expresses the belief that there are familiar circumstances when our senses are capable of seeing the world 'as it really is' and other, equally familiar circumstances, when they are deceived. It expresses the belief that some of our sensory impressions are veridical while others are not. We also take it for granted that the material medium between the object and us exerts no distorting influence, and that the physical entity that establishes the contact – light – carries a true picture. All these are abstract, and highly doubtful, assumptions which shape our view of the world without being accessible to a direct criticism. Usually, we are not even aware of them and we recognize their effects only when we encounter an entirely different cosmology: prejudices are found by contrast, not by analysis. The material which the *scientist* has at his disposal, his most sublime theories and his most sophisticated techniques included, is structured in exactly the same way. It again contains principles which are not known and which, if known, would be extremely hard to test. (As a result, a theory may clash with the evidence not because it is not correct, but because the evidence is contaminated.)

Now – how can we possibly examine something we are using all the time? How can we analyse the terms in which we habitually express our most simple and straightforward observations, and reveal their presuppositions? How can we discover the kind of world we presuppose when proceeding as we do?

The answer is clear: we cannot discover it from the *inside*. We need an *external* standard of criticism, we need a set of alternative assumptions or, as these assumptions will be quite general, constituting, as it were, an entire alternative world, *we need a dream-world in order to discover the features of the real world we think we inhabit* (and which may actually be just another dream-world). The first step in our criticism of familiar concepts and procedures, the first step in our criticism of 'facts', must therefore be an attempt to break the circle. We must invent a new conceptual system that suspends, or clashes with, the most carefully established observational results, confounds the most plausible theoretical principles, and introduces perceptions that cannot form part of the existing perceptual world.[1] This step is again counterinductive.

---

1. 'Clashes' or 'suspends' is meant to be more general than 'contradicts'. I shall say that a set of ideas or actions 'clashes' with a conceptual system if it is either inconsistent with it, or makes the system appear absurd. For details see Chapter 16 below.

Counterinduction is, therefore, always reasonable and it has always a chance of success.

In the following seven chapters, this conclusion will be developed in greater detail and it will be elucidated with the help of historical examples. One might therefore get the impression that I recommend a new methodology which replaces induction by counterinduction and uses a multiplicity of theories, metaphysical views, fairy-tales instead of the customary pair theory/observation.[2] This impression would certainly be mistaken. My intention is not to replace one set of general rules by another such set: my intention is, rather, to convince the reader that *all methodologies, even the most obvious ones, have their limits.* The best way to show this is to demonstrate the limits and even the irrationality of some rules which she, or he, is likely to regard as basic. In the case of induction (including induction by falsification) this means demonstrating how well the counterinductive procedure can be supported by argument. Always remember that the demonstrations and the rhetorics used do not express any 'deep convictions' of mine. They merely show how easy it is to lead people by the nose in a rational way. An anarchist is like an undercover agent who plays the game of Reason in order to undercut the authority of Reason (Truth, Honesty, Justice, and so on).[3]

---

2. This is how Professor Ernan McMullin interpreted some earlier papers of mine. See 'A Taxonomy of the Relations between History and Philosophy of Science', *Minnesota Studies*, Vol. 5, Minneapolis, 1971.

3. 'Dada', says Hans Richter in *Dada: Art and Anti-Art*, 'not only had no programme, it was against all programmes.' This does not exclude the skilful defence of programmes to show the chimerical character of any defence, however 'rational'. (In the same way an actor or a playwright could produce all the outer manifestations of 'deep love' in order to debunk the idea of 'deep love' itself. Example: Pirandello.)

3

> *The consistency condition which demands that new hypotheses agree with accepted theories is unreasonable because it preserves the older theory, and not the better theory. Hypotheses contradicting well-confirmed theories give us evidence that cannot be obtained in any other way. Proliferation of theories is beneficial for science, while uniformity impairs its critical power. Uniformity also endangers the free development of the individual.*

In this chapter I shall present more detailed arguments for the 'counterrule' that urges us to introduce hypotheses which are *inconsistent* with well-established *theories*. The arguments will be indirect. They will start with a criticism of the demand that new hypotheses must be consistent with such theories. This demand will be called the *consistency condition*.[1]

Prima facie, the case of the consistency condition can be dealt with in a few words. It is well known (and has also been shown in detail by Duhem) that Newton's mechanics is inconsistent with Galileo's law of free fall and with Kepler's laws; that statistical thermodynamics is inconsistent with the second law of the phenomenological theory; that wave optics is inconsistent with geometrical optics; and so on.[2] Note that what is being asserted here is *logical* inconsistency; it may well be that the differences of prediction are too small to be detected by experiment. Note also that what is being asserted is not the *inconsistency* of, say, Newton's *theory* and Galileo's law, but rather the inconsistency of *some consequences* of Newton's theory in the domain of validity of Galileo's law, and Galileo's law. In the last case, the situation is especially clear. Galileo's law asserts that the acceleration of free fall is a constant, whereas application of Newton's theory to the surface of the earth gives an acceleration that is not constant but *decreases* (although imperceptibly) with the distance from the centre of the earth.

To speak more abstractly: consider a theory T' that successfully describes the situation inside domain D'. T' agrees with a *finite* number

---

1. The consistency condition goes back to Aristotle at least. It plays an important part in Newton's philosophy (though Newton himself constantly violated it). It is taken for granted by many 20th-century scientists and philosophers of science.

2. Pierre Duhem, *The Aim and Structure of Physical Theory*, New York, 1962, pp. 180ff.

of observations (let their class be F) and it agrees with these observations inside a margin M of error. Any alternative that contradicts T′ outside F and inside M is supported by exactly the same observations and is therefore acceptable if T′ was acceptable (I shall assume that F are the only observations made). The consistency condition is much less tolerant. It eliminates a theory or a hypothesis not because it disagrees with the facts; it eliminates it because it disagrees with another theory, with a theory, moreover, whose confirming instances it shares. It thereby makes the as yet untested part of that theory a measure of validity. The only difference between such a measure and a more recent theory is age and familiarity. Had the younger theory been there first, then the consistency condition would have worked in its favour. 'The *first* adequate theory has the right of priority over equally adequate aftercomers.'[3] In this respect the effect of the consistency condition is rather similar to the effect of the more traditional methods of transcendental deduction, analysis of essences, phenomenological analysis, linguistic analysis. It contributes to the preservation of the old and familiar not because of any inherent advantage in it but because it is old and familiar. This is not the only instance where on closer inspection a rather surprising similarity emerges between modern empiricism and some of the school philosophies it attacks.

Now it seems to me that these brief considerations, although leading to an interesting *tactical* criticism of the consistency condition, and to some first shreds of support for counterinduction, do not yet go to the heart of the matter. They show that an alternative to the accepted point of view which shares its confirming instances cannot be *eliminated* by factual reasoning. They do not show that such an alternative is *acceptable*; and even less do they show that it *should be used*. It is bad enough, a defender of the consistency condition might point out, that the accepted view does not possess full empirical support. Adding new theories of *an equally unsatisfactory character* will not improve the situation; nor is there much sense in trying to *replace* the accepted theories by some of their possible alternatives. Such replacement will be no easy matter. A new formalism may have to be learned and familiar problems may have to be calculated in a new way. Textbooks must be rewritten, university curricula readjusted,

---

3. C. Truesdell, 'A Program Toward Rediscovering the Rational Mechanics of the Age of Reason', *Archives for the History of Exact Sciences*, Vol. 1, p. 14.

experimental results reinterpreted. And what will be the result of all the effort? Another theory which from an empirical standpoint has no advantage whatsoever over and above the theory it replaces. The only real improvement, so the defender of the consistency condition will continue, derives from the *addition of new facts*. Such new facts will either support the current theories, or they will force us to modify them by indicating precisely where they go wrong. In both cases they will precipitate real progress and not merely arbitrary change. The proper procedure must therefore consist in the confrontation of the accepted point of view with as many relevant facts as possible. The exclusion of alternatives is then simply a measure of expediency: their invention not only does not help, it even hinders progress by absorbing time and manpower that could be devoted to better things. The consistency condition eliminates such fruitless discussion and it forces the scientist to concentrate on the facts which, after all, are the only acceptable judges of a theory. This is how the practising scientist will defend his concentration on a single theory to the exclusion of empirically possible alternatives.

It is worthwhile repeating the reasonable core of this argument. Theories should not be changed unless there are pressing reasons for doing so. The only pressing reason for changing a theory is disagreement with facts. Discussion of incompatible facts will therefore lead to progress. Discussion of incompatible hypotheses will not. Hence, it is sound procedure to increase the number of relevant facts. It is not sound procedure to increase the number of factually adequate, but incompatible, alternatives. One might wish to add that formal improvements such as increased elegance, simplicity, generality, and coherence should not be excluded. But once these improvements have been carried out, the collection of facts for the purpose of tests seems indeed to be the only thing left to the scientist.

And so it is – provided facts *exist, and are available independently of whether or not one considers alternatives to the theory to be tested*. This assumption, on which the validity of the foregoing argument depends in a most decisive manner, I shall call the assumption of the relative autonomy of facts, or the *autonomy principle*. It is not asserted by this principle that the discovery and description of facts is independent of *all* theorizing. But it is asserted that the facts which belong to the empirical content of some theory are available whether or not one considers alternatives to *this* theory. I am not aware that this very important assumption has ever been explicitly formulated as a separate postulate of the empirical method. However, it is clearly implied in almost all investigations which deal with questions of confirmation and test. All these investigations use a model in which a *single* theory is compared with a class of facts (or observation statements) which are assumed to be 'given' somehow. I submit that this is much too simple a picture of the actual situation. Facts and theories are much more intimately connected than is admitted by the autonomy principle. Not only is the description of every single fact dependent on *some* theory (which may, of course, be very different from the theory to be tested), but there also exist facts which cannot be unearthed except with the help of alternatives to the theory to be tested, and which become unavailable as soon as such alternatives are excluded. This suggests that the methodological unit to which we must refer when discussing questions of test and empirical content is constituted by a *whole set of partly overlapping, factually adequate, but mutually inconsistent theories.* In the present chapter only the barest outlines will be given of such a test model. However, before doing this, I want to discuss an example which shows very clearly the function of alternatives in the discovery of critical facts.

It is now known that the Brownian particle is a perpetual motion machine of the second kind and that its existence refutes the phenomenological second law. Brownian motion therefore belongs to the domain of relevant facts for the law. Now could this relation between Brownian motion and the law have been discovered in a *direct* manner, i.e. could it have been discovered by an examination of the observational consequences of the phenomenological theory that did not make use of an alternative theory of heat? This question is readily divided into two: (1) Could the *relevance* of the Brownian particle have been discovered in this manner? (2) Could it have been demonstrated that it actually *refutes* the second law?

The answer to the first question is that we do not know. It is impossible to say what would have happened if the kinetic theory had not been introduced into the debate. It is my guess, however, that in that case the Brownian particle would have been regarded as an oddity – in much the same way as some of the late Professor Ehrenhaft's astounding effects were regarded as an oddity, and that it would not have been given the decisive position it assumed in contemporary theory. The answer to the second question is simply – No. Consider what the discovery of an inconsistency between the phenomenon of Brownian motion and the second law would have required. It would have required: (a) measurement of the exact *motion* of the particle in order to ascertain the change in its kinetic energy plus the energy spent on overcoming the resistance of the fluid; and (b) precise measurements of temperature and heat transfer in the surrounding medium in order to establish that any loss occurring there was indeed compensated by the increase in the energy of the moving particle and the work done against the fluid. Such measurements are beyond experimental possibilities;[4] neither the heat transfer nor the path of the particle can be measured with the desired precision. Hence a 'direct' refutation of the second law that considers only the phenomenological theory and the 'facts' of the Brownian motion is impossible. It is impossible because of the structure of the world in which we live and because of the laws that are valid in this world. And as is well known, the actual refutation was brought about in a very different manner. It was brought about via the kinetic theory and Einstein's utilization of it in his calculation of the statistical properties of Brownian motion. In the course of this procedure, the phenomenological theory (T') was incorporated into the wider context of statistical physics (T) in such a manner that *the consistency condition was violated*, and it was only *then* that crucial experiments were staged (investigations of Svedberg and Perrin).[5]

---

4. For details see R. Fürth, *Zs. Physik*, Vol. 81, 1933, pp. 143ff.

5. For these investigations (whose philosophical background derives from Boltzmann) see A. Einstein, *Investigation on the Theory of the Brownian Motion*, ed. R. Fürth, New York, 1956, which contains all the relevant papers by Einstein and an exhaustive bibliography by R. Fürth. For the experimental work of J. Perrin, see *Die Atome*, Leipzig, 1920. For the relation between the phenomenological theory and the kinetic theory of von Smoluchowski, see 'Experimentell nachweisbare, der üblichen Thermodynamik widersprechende Molekularphänomene', *Physikalische Zs.*, Vol. 8, 1912, p. 1069, as well as the brief note by K.R. Popper, 'Irreversibility, or, Entropy since 1905', *British Journal for the Philosophy of Science*, Vol. 8, 1957, p. 151, which summarizes the essential arguments. Despite Einstein's epoch-making discoveries and von Smoluchowski's splendid presentation of their consequences (*Oeuvres de Marie Smoluchowski*, Cracow, 1927, Vol. 2, pp. 226ff, 316ff, 462ff and 530ff), the present situation in thermodynamics is extremely unclear, especially in view of the continued presence of some very doubtful ideas of reduction. To be more specific, the attempt is frequently made to determine the entropy balance of a complex *statistical* process by reference to the (refuted) *phenomenological* law after which fluctuations are inserted in an *ad hoc* fashion. For this see my note 'On the Possibility of a Perpetuum Mobile of the Second Kind', *Mind, Matter and Method*, Minneapolis, 1966, p. 409, and my paper 'In Defence of Classical Physics', *Studies in the History and Philosophy of Science* 1, No. 2, 1970.

It seems to me that this example is typical of the relation between fairly general theories, or points of view, and the 'facts'. Both the relevance and the refuting character of decisive facts can be established only with the help of other theories which, though factually adequate,[6] are not in agreement with the view to be tested. This being the case, the invention and articulation of alternatives may have to precede the production of refuting facts. Empiricism, at least in some of its more sophisticated versions, demands that the empirical content of whatever knowledge we possess be increased as much as possible. *Hence the invention of alternatives to the view at the centre of discussion constitutes an essential part of the empirical method.* Conversely the fact that the consistency condition eliminates alternatives now shows it to be in disagreement not only with scientific practice but with empiricism as well. By excluding valuable tests it decreases the empirical content of the theories that are permitted to remain (and these, as I have indicated above, will usually be the theories which were there first); and it especially decreases the number of those facts that could show their limitations. This is how empiricists (such as Newton, or some proponents of what has been called the orthodox interpretation of quantum mechanics) who defend the consistency condition, being unaware of the complex nature of scientific knowledge (and, for that matter, of any form of knowledge) are voiding their favourite theories of empirical content and thus turning them into what they most despise, viz. metaphysical doctrines.[7]

It ought to be mentioned, incidentally, that in 1903, when Einstein started his work in thermodynamics, there existed empirical evidence suggesting that Brownian motion could not be a molecular phenomenon. See F.M. Exner, 'Notiz zu Browns Molekularbewegung', *Ann. Phys.*, No. 2, 1900, p. 843. Exner claimed that the motion was of orders of magnitude beneath the value to be expected on the equipartition principle. Einstein (*Investigations in the Theory of the Brownian Movement*, pp. 63ff, esp. p. 67) gave the following theoretical explanation of the discrepancy: 'since an observer operating with definite means of observation in a definite manner can never perceive the actual path transversed in an arbitrarily small time, a certain mean velocity will always appear to him as an instantaneous velocity. But it is clear that the velocity ascertained thus corresponds to no objective property of the motion under investigation.' See also Mary Jo Nye, *Molecular Reality*, London, 1972, pp. 98ff.

6. The condition of factual adequacy will be removed in Chapter 5.

7. The most dramatic confirmation of the orthodox view which made its empirical nature obvious came by way of Bell's theorem. But Bell was on the side of Einstein, not of Bohr, whom he regarded as an 'obscurantist'. See Jeremy Bernstein, *Quantum Profiles*, Princeton, 1991, pp. 3ff (for Bell's background) and p. 84 (for 'obscurantist').

John Stuart Mill has given a fascinating account of the gradual transformation of revolutionary ideas into obstacles to thought. When a new view is proposed it faces a hostile audience and excellent reasons are needed to gain for it an even moderately fair hearing. The reasons are produced, but they are often disregarded or laughed out of court, and unhappiness is the fate of the bold inventors. But new generations, being interested in new things, become curious; they consider the reasons, pursue them further and groups of researchers initiate detailed studies. The studies may lead to surprising successes (they also raise lots of difficulties). Now nothing succeeds like success, even if it is success surrounded by difficulties. The theory becomes acceptable as a topic for discussion; it is presented at meetings and large conferences. The diehards of the status quo feel an obligation to study one paper or another, to make a few grumbling comments, and perhaps to join in its exploration. There comes then a moment when the theory is no longer an esoteric discussion topic for advanced seminars and conferences, but enters the public domain. There are introductory texts, popularizations; examination questions start dealing with problems to be solved in its terms. Scientists from distant fields and philosophers, trying to show off, drop a hint here and there, and this often quite uninformed desire to be on the right side is taken as a further sign of the importance of the theory.

Unfortunately, this increase in importance is not accompanied by better understanding; the very opposite is the case. Problematic aspects which were originally introduced with the help of carefully constructed arguments now become basic principles; doubtful points turn into slogans; debates with opponents become standardized and also quite unrealistic, for the opponents, having to express themselves in terms which presuppose what they contest, seem to raise quibbles, or to misuse words. Alternatives are still employed but they no longer contain realistic counter-proposals; they only serve as a background for the splendour of the new theory. Thus we do have success – but it is the success of a manoeuvre carried out in a void, overcoming difficulties that were set up in advance for easy solution. An empirical theory such as quantum mechanics or a pseudo-empirical practice such as modern scientific medicine with its materialistic background can of course point to numerous achievements but *any* view and *any* practice that has been around for some time has achievements. The question is whose achievements are better or more important, and *this* question cannot be answered for there are no realistic alternatives to provide a point of comparison. A wonderful invention has turned into a fossil.

There exist numerous historical examples of the process I have just described, and various authors have commented on it. The most important recent author is Professor Thomas Kuhn. In his *Structure of Scientific Revolutions*,[8] he distinguishes between science and pre-science and, within science, between revolutions and normal science. Pre-science, according to him, is pluralistic throughout and therefore in danger of concentrating on opinions rather than on things (Bacon made a similar point). The two components of mature science perfectly agree with the two stages mentioned above except that Kuhn doubts that science or, for that matter, any activity that claims to produce factual knowledge can do without a normal component. Fossils, he seems to say, are needed to give substance to the debates that occur in the revolutionary component – but he adds that the latter cannot advance without alternatives. Two earlier authors are Mill and Niels Bohr. Mill gives a clear and compelling description of the transition from the early stage of a new view to its orthodoxy. Debates and reasoning, he writes, are features

> belonging to periods of transition, when old notions and feelings have been unsettled and no new doctrines have yet succeeded to their ascendancy. At such times people of any mental activity, having given up their old beliefs, and not feeling quite sure that those they still retain can stand unmodified listen eagerly to new opinions. But this state of things is necessarily transitory: some particular body of doctrine in time rallies the majority round it, organizes social institutions and modes of action conformably to itself, education impresses this new creed upon the new generation *without the mental processes that have led to it* and by degrees it acquires the very same power of compression, so long exercised by the creeds of which it had taken the place.[9]

An account of the alternatives replaced, of the process of replacement, of the arguments used in its course, of the strength of the old views and the weaknesses of the new, not a 'systematic account' but a *historical account*

---

8. Chicago, 1962.

9. 'Autobiography', quoted from *Essential Works of John Stuart Mill*, ed. M. Lerner, New York, 1965, p. 119; my emphasis.

of each stage of knowledge, can alleviate these drawbacks and increase the rationality of one's theoretical commitments. Bohr's presentation of new discoveries has precisely this pattern; it contains preliminary summaries surveying the past, moves on to the 'present state of knowledge' and ends up by making general suggestions for the future.[10]

Mill's views and Bohr's procedure are not only an expression of their liberal attitude; they also reflect their conviction that a pluralism of ideas and forms of life is an essential part of any rational inquiry concerning the nature of things. Or, to speak more generally: *Unanimity of opinion may be fitting for a rigid church, for the frightened or greedy victims of some (ancient, or modern) myth, or for the weak and willing followers of some tyrant. Variety of opinion is necessary for objective knowledge. And a method that encourages variety is also the only method that is compatible with a humanitarian outlook.* (To the extent to which the consistency condition delimits variety, it contains a theological element which lies, of course, in the worship of 'facts' so characteristic of nearly all empiricism.[11])

---

10. For a more detailed account see my *Philosophical Papers*, Vol. 1, Chapter 16, section 6.

11. It is interesting to see that the platitudes that directed the Protestants to the Bible are often almost identical with the platitudes which direct empiricists and other fundamentalists to *their* foundation, viz. experience. Thus in his *Novum Organum* Bacon demands that all preconceived notions (aphorism 36), opinions (aphorisms 42ff), even *words* (aphorisms 59, 121), 'be adjured and renounced with firm and solemn resolution, and the understanding must be completely freed and cleared of them, so that the access to the kingdom of man, which is founded on the sciences, may resemble that to the kingdom of heaven, where no admission is conceded, except to children' (aphorism 68). In both cases 'disputation' (which is the consideration of alternatives) is criticized, in both cases we are invited to dispense with it, and in both cases we are promised an 'immediate perception', here, of God, and there of Nature. For the theoretical background of this similarity see my essay 'Classical Empiricism', in R.E. Butts (ed.), *The Methodological Heritage of Newton*, Oxford and Toronto, 1970. For the strong connections between Puritanism and modern science see R.T. Jones, *Ancients and Moderns*, California, 1965, Chapters 5–7. A thorough examination of the factors that influenced the rise of modern empiricism in England is found in R.K. Merton, *Science, Technology and Society in Seventeenth Century England*, New York, 1970 (book version of the 1938 article).

4

There is no idea, however ancient and absurd, that is not capable of improving our knowledge. The whole history of thought is absorbed into science and is used for improving every single theory. Nor is political interference rejected. It may be needed to overcome the chauvinism of science that resists alternatives to the status quo.

This finishes the discussion of part one of counterinduction dealing with the invention and elaboration of hypotheses inconsistent with a point of view that is highly confirmed and generally accepted. The result was that a thorough examination of such a point of view may involve incompatible alternatives so that the (Newtonian) advice to postpone alternatives until after the first difficulty has arisen means putting the cart before the horse. A scientist who is interested in maximal empirical content, and who wants to understand as many aspects of his theory as possible, will adopt a pluralistic methodology, he will compare theories with other theories rather than with 'experience', 'data', or 'facts', and he will try to improve rather than discard the views that appear to lose in the competition.[1] For the alternatives, which he needs to keep the contest going, may be taken from the past as well. As a matter of fact, they may be taken from wherever one is able to find them – from ancient myths and modern prejudices; from the lucubrations of experts and from the fantasies of cranks. The whole history of a subject is utilized in the attempt to improve its most recent and most 'advanced' stage. The separation between the history of a science, its philosophy and the science itself dissolves into thin air and so does the separation between science and non-science.[2]

---

1. It is, therefore, important that the alternatives be set against each other and not be isolated or emasculated by some form of 'demythologization'. Unlike Tillich, Bultmann and their followers, we should regard the world-views of the Bible, the Gilgamesh epic, the Iliad, the Edda, as fully fledged *alternative cosmologies* which can be used to modify, and even to replace, the 'scientific' cosmologies of a given period.

2. An account and a truly humanitarian defence of this position can be found in J.S. Mill's *On Liberty*. Popper's philosophy, which some people would like to lay on us as the one and only humanitarian rationalism in existence today, is but a pale reflection of Mill. It is specialized, formalistic and elitist, and devoid of the concern for individual happiness that is such a characteristic feature of Mill. We can understand its peculiarities when we consider (a) the background of logical positivism, which plays an important role in the

This position, which is a natural consequence of the arguments presented above, is frequently attacked – not by counter-arguments, which would be easy to answer, but by rhetorical questions. 'If any metaphysics goes,' writes Dr Hesse in her review of an earlier essay of mine,[3] 'then the question arises why we do not *go back* and exploit the objective criticism of modern science available in Aristotelianism, or indeed in Voodoo?' – and she insinuates that a criticism of this kind

---

Logic of Scientific Discovery, (b) the unrelenting puritanism of its author (and of most of his followers), and when we remember the influence of Harriet Taylor on Mill's life and on his philosophy. There is no Harriet Taylor in Popper's life. The foregoing arguments should also have made it clear that I regard proliferation not just as an 'external catalyst' of progress, as Lakatos suggests in his essays ('History of Science and Its Rational Reconstructions', *Boston Studies*, Vol. 8, p. 98; 'Popper on Demarcation and Induction', MS, 1970, p. 21), but as an essential part of it. Ever since 'Explanation, Reduction, and Empiricism' (*Minnesota Studies*, Vol. 3, Minneapolis, 1962), and especially in 'How to Be a Good Empiricist' (*Delaware Studies*, Vol. 2, 1963), I have argued that alternatives increase the empirical content of the views that happen to stand in the centre of attention and are, therefore, '*necessary* parts' of the falsifying process (Lakatos, 'History', fn. 27, describing his own position). In 'Reply to Criticism' (*Boston Studies*, Vol. 2, 1965) I pointed out that 'the principle of proliferation not only recommends invention of *new* alternatives, it also prevents the elimination of *older* theories which have been refuted. The reason is that such theories contribute to the content of their victorious rivals' (p. 224). This agrees with Lakatos' observation of 1971 that 'alternatives are not merely catalysts, which can later be removed in the rational reconstruction' ('History', fn. 27), *except* that Lakatos attributes the psychologistic view to me and my *actual* views to himself. Considering the argument in the text, it is clear that the increasing separation of the history, the philosophy of science and of science itself is a disadvantage and should be terminated in the interest of all these three disciplines. Otherwise we shall get tons of minute, precise, but utterly barren results.

3. Mary Hesse, *Ratio*, No. 9, 1967, p. 93; cf. B.F. Skinner, *Beyond Freedom and Dignity*, New York, 1971, p. 5: 'No modern physicist would turn to Aristotle for help.' This is neither true, nor would it be an advantage if it were true. Aristotelian ideas influenced research long after they had allegedly been removed by early modern astronomy and physics – any history of 17th- or 18th-century science will show that (example: John Heilbronn's marvellous *Electricity in the 17th and 18th Centuries*, Berkeley and Los Angeles, 1979). They resurfaced in biology, in the thermodynamics of open systems and even in mathematics. Aristotle's theory of locomotion (which has the consequence that a moving object has no precise length and that an object having a precise location must be at rest) was more advanced than the Galilean view and showed that ideas which in our time emerged from empirical research can be obtained by a careful analysis of the problems of the continuum (details on this point in Chapter 8 of my *Farewell to Reason*, London, 1987). Here as elsewhere the propagandists of a naive scientism give themselves the air of presenting arguments when all they do is spread unexamined and ill-conceived rumours.

would be altogether laughable. Her insinuation, unfortunately, assumes a great deal of ignorance in her readers. Progress was often achieved by a 'criticism from the past', of precisely the kind that is now dismissed by her. After Aristotle and Ptolemy, the idea that the earth moves – that strange, ancient, and 'entirely ridiculous',[4] Pythagorean view – was thrown on the rubbish heap of history, only to be revived by Copernicus and to be forged by him into a weapon for the defeat of its defeaters. The Hermetic writings played an important part in this revival, which is still not sufficiently understood,[5] and they were studied with care by the great Newton himself.[6] Such developments are not surprising. No idea is ever examined in all its ramifications and no view is ever given all the chances it deserves. Theories are abandoned and superseded by more fashionable accounts long before they have had an opportunity to show their virtues. Besides, ancient doctrines and 'primitive' myths appear strange and nonsensical only because the information they contain is either not known, or is distorted by philologists or anthropologists unfamiliar with the simplest physical, medical or astronomical knowledge.[7] Voodoo, Dr

---

4. Ptolemy, *Syntaxis*, quoted after the translation of Manitius, *Des Claudius Ptolemaeus Handbuch der Astronomie*, Vol. 1, Leipzig, 1963, p. 18.

5. For a positive evaluation of the role of the hermetic writings during the Renaissance cf. F. Yates, *Giordano Bruno and the Hermetic Tradition*, London, 1963, and the literature given there. For a criticism of her position see the articles by Mary Hesse and Edward Rosen in Vol. 5 of the *Minnesota Studies for the Philosophy of Science*, ed. Roger Stuewer, Minneapolis, 1970; R.S. Westman and J.E. McGuire, *Hermeticism and the Scientific Revolution*, Los Angeles, Clark Memorial Library, 1977, as well as Brian Vickers, *Journal of Modern History*, 51, 1979.

6. Cf. J.M. Keynes, 'Newton the Man', in Essays and Sketches in Biography, New York, 1956, and, in much greater detail, McGuire and Rattansi, 'Newton and the "Pipes of Pan"', *Notes and Records of the Royal Society*, Vol. 21, No. 2, 1966, pp. 108ff. For more detailed accounts cf. Frank Manuel, *The Religion of Isaac Newton*, Oxford, 1974, R.S. Westfall's monumental biography, *Never at Rest*, Cambridge, 1980, with literature, as well as Chapters X and XI of R. Popkin, *The Third Force in Seventeenth-Century Thought*, Leiden and New York, 1992.

7. For the scientific content of some myths see C. de Santillana, *The Origin of Scientific Thought*, New York, 1961, especially the Prologue. 'We can see then', writes de Santillana, 'how so many myths, fantastic and arbitrary in semblance, of which the Greek tale of the Argonauts is a late offspring, may provide a terminology of image motifs, a kind of code which is beginning to be broken. It was meant to allow those who knew (a) to determine unequivocally the position of given planets in respect to the earth, to the firmament, and to one another; (b) to present what knowledge there was of the fabric of the world in the

Hesse's *pièce de resistance*, is a case in point. Nobody knows it, everybody uses it as a paradigm of backwardness and confusion. And yet Voodoo has a firm though still not sufficiently understood material basis, and a study of its manifestations can be used to enrich, and perhaps even to revise, our knowledge of physiology.[8]

An even more interesting example is the revival of traditional medicine in Communist China. We start with a familiar development:[9] a great country with great traditions is subjected to Western domination and is exploited in the customary way. A new generation recognizes or thinks it recognizes the material and intellectual superiority of the West and traces it back to science. Science is imported, taught, and pushes aside all traditional elements. Scientific chauvinism triumphs: 'What is compatible with science should live, what is not compatible with science, should

---

form of tales about "how the world began". There are two reasons why this code was not discovered earlier. One is the firm conviction of historians of science that science did not start before Greece and that scientific results can only be obtained with the scientific method as it is practised today (and as it was foreshadowed by Greek scientists). The other reason is the astronomical, geological, etc., ignorance of most Assyriologists, Aegyptologists, Old Testament scholars, and so on: the apparent primitivism of many myths is just the reflection of the primitive astronomical, biological, etc., etc., knowledge of their collectors and translators. Since the discoveries of Hawkins, Marshack, Seidenberg, van der Waerden (*Geometry and Algebra in Ancient Civilizations*, New York, 1983) and others we have to admit the existence of an international palaeolithic astronomy that gave rise to schools, observatories, scientific traditions and most interesting theories. These theories, which were expressed in sociological, not in mathematical, terms, have left their traces in sagas, myths, legends, and may be reconstructed in a twofold way, by going *forward* into the present from the material remains of Stone Age astronomy such as marked stones, stone observatories, etc., and by going *back* into the past from the literary remains which we find in sagas, legends, myths. An example of the first method is A. Marshack, *The Roots of Civilization*, New York, 1972. An example of the second is de Santillana-von Dechend, *Hamlet's Mill*, Boston, 1969.

8. See Chapter 9 of Lévi-Strauss, *Structural Anthropology*, New York, 1967. For the physiological basis of Voodoo see C.R. Richter, 'The Phenomenon of Unexplained Sudden Death', in Gantt (ed.), *The Physiological Basis of Psychiatry*, as well as W.H. Cannon, *Bodily Changes in Pain, Hunger, Fear and Rage*, New York, 1915, and ' "Voodoo" Death', in *American Anthropologist*, n.s., XLIV, 1942. The detailed biological and meteorological observations made by so-called 'primitives' are reported in Lévi-Strauss, *The Savage Mind*, London, 1966.

9. R.C. Croizier, *Traditional Medicine in Modern China*, Cambridge, Mass., 1968. The author gives a very interesting and fair account of developments with numerous quotations from newspapers, books, pamphlets, but is often inhibited by his respect for 20th-century science.

die'.[10] 'Science' in this context means not just a specific method, but all the results the method has so far produced. Things incompatible with the results must be eliminated. Old-style doctors, for example, must either be removed from medical practice, or they must be re-educated. Herbal medicine, acupuncture, moxibustion and the underlying philosophy are a thing of the past, no longer to be taken seriously. This was the attitude up to about 1954, when the condemnation of bourgeois elements in the Ministry of Health started a campaign for the revival of traditional medicine. No doubt the campaign was politically inspired. It contained at least two elements, viz. (1) the identification of Western science with bourgeois science and (2) the refusal of the party to exempt science from political supervision[11] and to grant experts special privileges. But it provided the counterforce that was needed to overcome the scientific chauvinism of the time and to make a plurality (actually a duality) of views possible. (This is an important point. It often happens that parts of science become hardened and intolerant so that proliferation must be enforced from the outside, and by political means. Of course, success cannot be guaranteed – see the Lysenko affair. But this does not remove the need for non-scientific controls on science.)

Now this politically enforced dualism has led to most interesting and puzzling discoveries both in China and in the West and to the realization that there are effects and means of diagnosis which modern medicine cannot repeat and for which it has no explanation. It revealed sizeable lacunae in Western medicine. Nor can one expect that the customary scientific approach will eventually find an answer. In the case of herbal medicine the approach consists of two steps.[12] First, the herbal concoction is analysed into its chemical constituents. Then the *specific* effects of each constituent are determined and the total effect on a particular organ explained on their basis. This neglects the possibility that the herb, taken in its entirety, changes the state of the *whole* organism and that it is this new state of the whole organism rather than a specific part of the herbal concoction, a 'magic bullet', as it were, that cures the diseased organ.

---

10. Chou Shao, 1933, as quoted in Croizier, op. cit., p. 109. See also D.W.Y. Kwok, *Scientism in Chinese Thought*, New Haven, 1965.

11. For the tensions between 'red' and 'expert' see F. Schurmann, *Ideology and Organization in Communist China*, Berkeley, 1966.

12. See M.B. Krieg, *Green Medicine*, New York, 1964.

Here as elsewhere knowledge is obtained from a multiplicity of views rather than from the determined application of a preferred ideology. And we realize that proliferation may have to be enforced by non-scientific agencies whose power is sufficient to overcome the most powerful scientific institutions. Examples are the Church, the State, a political party, public discontent, or money: the best single entity to get a modern scientist away from what his 'scientific conscience' tells him to pursue is still the *dollar* (or, more recently, the Swiss franc).

Pluralism of theories and metaphysical views is not only important for methodology, it is also an essential part of a humanitarian outlook. Progressive educators have always tried to develop the individuality of their pupils and to bring to fruition the particular, and sometimes quite unique, talents and beliefs of a child. Such an education, however, has very often seemed to be a futile exercise in day-dreaming. For is it not necessary to prepare the young for life *as it actually is*? Does this not mean that they must learn *one particular set of views* to the exclusion of everything else? And, if a trace of their imagination is still to remain, will it not find its proper application in the arts or in a thin domain of dreams that has but little to do with the world we live in? Will this procedure not finally lead to a split between a hated reality and welcome fantasies, science and the arts, careful description and unrestrained self-expression? The argument for proliferation shows that this need not happen. It is possible to *retain* what one might call the freedom of artistic creation *and to use it to the full*, not just as a road of escape but as a necessary means for discovering and perhaps even changing the features of the world we live in. This coincidence of the part (individual man) with the whole (the world we live in), of the purely subjective and arbitrary with the objective and lawful, is one of the most important arguments in favour of a pluralistic methodology. For details the reader is advised to consult Mill's magnificent essay *On Liberty*.[13]

---

13. See my account of this essay in Vol. 1, Chapter 8 and Vol. 2, Chapter 4 of my *Philosophical Papers*. See also Appendix 1 of the present essay.

5

No theory ever agrees with all the facts in its domain, yet it is not always the theory that is to blame. Facts are constituted by older ideologies, and a clash between facts and theories may be proof of progress. It is also a first step in our attempt to find the principles implicit in familiar observational notions.

Considering now the invention, elaboration and the use of theories which are inconsistent, not just with other theories, but even with *experiments, facts, observations,* we may start by pointing out that *no single theory ever agrees with all the known facts in its domain.* And the trouble is not created by rumours, or by the result of sloppy procedure. It is created by experiments and measurements of the highest precision and reliability.

It will be convenient, at this place, to distinguish two different kinds of disagreement between theory and fact: numerical disagreement, and qualitative failures.

The first case is quite familiar: a theory makes a certain numerical prediction and the value that is actually obtained differs from the prediction made by more than the margin of error. Precision instruments are usually involved here. Numerical disagreements abound in science. They give rise to an 'ocean of anomalies' that surrounds every single theory.[1]

Thus the Copernican view at the time of Galileo was inconsistent with facts so plain and obvious that Galileo had to call it 'surely false'.[2] 'There is no limit to my astonishment,' he writes in a later work,[3] 'when I reflect that Aristarchus and Copernicus were able to make reason so conquer sense that, in defiance

---

1. For the 'ocean' and various ways of dealing with it, see my 'Reply to Criticism', *Boston Studies*, Vol. 2, 1965, pp. 224ff.

2. Galileo Galilei, *The Assayer*, quoted in S. Drake and C.D. O'Malley (eds), *The Controversy on the Comets of 1618*, London, 1960, p. 185. The 'surely false' refers to the condemnation by Church authorities. But, as will be explained in the course of the book and especially in Chapter 13, the condemnation was based in part on the 'philosophical absurdity' of the idea of a moving earth, i.e. on its empirical failures and its theoretical inadequacy. See also the next quotation and footnote. 'As to the system of Ptolemy', writes Galileo on this point (p. 184), 'neither Tycho, nor other astronomers, nor even Copernicus could clearly refute it, inasmuch as a most important argument taken from the movement of Mars and Venus always stood in their way.' The 'most important argument' and Galileo's resolution are discussed in Chapters 9 and 10.

3. Galileo Galilei, *Dialogue Concerning the Two Chief World Systems*, Berkeley, 1953, p. 328.

of the latter, the former became mistress of their belief. Newton's theory of gravitation was beset, from the very beginning, by difficulties serious enough to provide material for refutation.[4] Even quite recently and in the non-relativistic domain it could be said that there 'exist numerous discrepancies between observation and theory.'[5] Bohr's atomic model was introduced, and retained, in the face of precise and unshakeable contrary evidence.[6] The special theory of relativity was retained despite Kaufmann's unambiguous results of 1906, and despite D.C. Miller's experiment.[7] The general theory

---

4. According to Newton the 'mutual actions of comets and planets upon one another' give rise to 'some inconsiderable irregularities . . . which will be apt to increase, till the system wants a reformation', *Opticks*, New York, 1952, p. 402. What Newton means is that gravitation disturbs the planets in a way that is likely to blow the planetary system apart. Babylonian data as used by Ptolemy show that the planetary system has remained stable for a long time. Newton concluded that it was being periodically 'reformed' by divine interventions: God acts as a stabilizing force in the planetary system (and in the world as a whole, which is constantly losing motion through processes such as inelastic collisions). One of the 'irregularities' considered by Newton, the great inequality of Jupiter and Saturn (*Principia*, transl. Motte, ed. Cajori, Berkeley, 1934, p. 397) was shown by Laplace to be a periodic disturbance with a large period. Then Poincaré found that the series developments customary in the calculations often diverged after they had shown some convergence, while Bruhns discovered that no quantitative methods other than series expansions could resolve the n-body problem. This was the end of the purely quantitative period in celestial mechanics (details in J. Moser, *Annals of Mathematical Studies*, Vol. 77, 1973, Princeton). See also M. Ryabov, *An Elementary Survey of Celestial Mechanics*, New York, 1961, for a survey and quantitative results of various methods of calculation. The qualitative approach is briefly described on pp. 126f. Thus it took more than two hundred years before one of the many difficulties of this rather successful theory was finally resolved.

5. Brower-Clemence, *Method of Celestial Mechanics*, New York, 1961. Also R.H. Dicke, 'Remarks on the Observational Basis of General Relativity', in H.Y. Chiu and W.F. Hoffman (eds), *Gravitation and Relativity*, New York, 1964, pp. 1–16. For a more detailed discussion of some of the difficulties of classical celestial mechanics, cf. J. Chazy, *La Théorie de la relativité et la Méchanique céleste*, Vol. 1, Chapters 4 and 5, Paris, 1928.

6. See Max Jammer, *The Conceptual Development of Quantum Mechanics*, New York, 1966, section 22. For an analysis see section 3c/2 of Lakatos, 'Falsification and the Methodology of Scientific Research Programmes', in Lakatos and Musgrave (eds), *Criticism and the Growth of Knowledge*, Cambridge, 1970.

7. W. Kaufmann, 'Über die Konstitution des Elektrons', *Ann. Phys.*, No. 19, 1906, p. 487. Kaufmann stated his conclusion quite unambiguously, and in italics: '*The results of the measurements are not compatible with the fundamental assumption of Lorentz and Einstein.*' Lorentz's reaction: '. . . it seems very likely that we shall have to relinquish this idea altogether' (*Theory of Electrons*, second edition, p. 213). Ehrenfest: 'Kaufmann demonstrates that Lorentz's deformable electron is ruled out by the measurements' ('Zur Stabilitätsfrage bei

of relativity, though surprisingly successful in a series of occasionally rather dramatic tests,[8] had a rough time in areas of celestial mechanics different from the advance of the perihelion of Mercury.[9] In the sixties the arguments and observations of Dicke and others seemed to endanger even this prediction. The problem is still unresolved.[10] On the other hand there

---

den Bucherer–Langevin Elektronen', *Phys. Zs.*, Vol. 7, 1906, p. 302). Poincaré's reluctance to accept the 'new mechanics' of Lorentz can be explained, at least in part, by the outcome of Kaufmann's experiment. See *Science and Method*, New York 1960, Book III, Chapter 2, section V, where Kaufmann's experiment is discussed, the conclusion being that the 'principle of relativity . . . cannot have the fundamental importance one was inclined to ascribe to it'. See also St Goldberg, 'Poincaré's Silence and Einstein's Relativity', *British Journal for the History of Science*, Vol. 5, 1970, pp. 73ff, and the literature given there. Einstein alone regarded the results as 'improbable because their basic assumptions, from which the mass of the moving electron is deduced, are not suggested by theoretical systems which encompass wider complexes of phenomena' (*Jahrbuch der Radioaktivität und Elektrizität*, Vol. 4, 1907, p. 349). Miller's work was studied by Lorentz for many years, but he could not find the trouble. It was only in 1955, twenty-five years after Miller had finished his experiments, that a satisfactory account of Miller's results was found. See R.S. Shankland, 'Conversations with Einstein', *Am. Journ. Phys.*, Vol. 31, 1963, pp. 47–57, especially p. 51, as well as footnotes 19 and 34; see also the inconclusive discussion at the 'Conference on the Michelson–Morley Experiment', *Astrophysical Journal*, Vol. 68, 1928, pp. 341ff.

Kaufmann's experiment was analysed by Max Planck and found to be not decisive: what had stopped Ehrenfest, Poincaré and Lorentz did not stop Planck. Why? My conjecture is that Planck's firm belief in an objective reality and his assumption that Einstein's theory was about such a reality made him a little more critical. Details in Chapter 6 of Elie Zahar, *Einstein's Revolution*, La Salle, Ill., 1989.

8. Such as the test of the effects of gravity upon light that was carried out in 1919 by Eddington and Crommelin and evaluated by Eddington. For a colourful description of the event and its impact, see C.M. Will, *Was Einstein Right?*, New York, 1986, pp. 75ff.

9. Chazy, op. cit., p. 230.

10. Repeating considerations by Newcomb (reported, for example, in Chazy, op. cit., pp. 204ff), Dicke pointed out that an oblateness of the sun would add classical terms to Mercury's motion and reduce the excess (compared with Newton's theory) advance of its perihelion. Measurements by Dicke and Goldenberg then found a difference of 52 km between the equatorial and polar diameter of the sun and a corresponding reduction of three seconds of arc for Mercury – a sizeable deviation from the relativistic value. This led to a considerable controversy concerning the accuracy of the Dicke–Goldenberg experiment and to an increase in the number of non-Einsteinian theories of gravitation. Technical details in C.M. Will, *Theory and Experiment in Gravitational Physics*, Cambridge, 1981, pp. 176ff, a popular survey including later developments in *Was Einstein Right?*, Chapter 5. Note how a new theory (Einstein's theory of gravitation) which is theoretically plausible and well confirmed can be endangered by exploiting its 'refuted' predecessor and carrying out appropriate experiments. See also Dicke, op. cit.

exist numerous new tests, both inside the planetary system and outside of it,[11] that provide confirmations of a precision unheard of only twenty years ago and unimagined by Einstein. In most of these cases we are dealing with quantitative problems which can be resolved by discovering a better set of *numbers* but which do not force us to make qualitative adjustments.[12]

---

11. Tests outside the planetary system (cosmology, black holes, pulsars) are needed to examine alternatives that agree with Einsteinian relativity inside the solar system. There now exists a considerable number of such alternatives and special steps have been taken to classify them and to elucidate their similarities and differences. See the introduction to Will, op. cit.

12. The situation just described shows how silly it would be to approach science from a naive-falsificationist perspective. Yet this is precisely what some philosophers have been trying to do. Thus Herbert Feigl (*Minnesota Studies*, Vol. 5, 1971, p. 7) and Karl Popper (*Objective Knowledge*, p. 78) have tried to turn Einstein into a naive falsificationist. Feigl writes: 'If Einstein relied on "beauty", "harmony", "symmetry", "elegance" in constructing . . . his general theory of relativity, it must nevertheless be remembered that he also said (in a lecture in Prague in 1920 – I was present then as a very young student): "If the observations of the red shift in the spectra of massive stars don't come out quantitatively in accordance with the principles of general relativity, then my theory will be dust and ashes". Popper writes: 'Einstein . . . said that if the red shift effect . . . was not observed in the case of white dwarfs, his theory of general relativity would be refuted.'

Popper gives no source for his story, and he most likely has it from Feigl. But Feigl's story and Popper's repetition conflict with the numerous occasions where Einstein emphasizes the 'reason of the matter' ('die Vernunft der Sache') over and above 'verification of little effects,' and this not only in casual remarks during a lecture, but in writing. Cf. the quotation in footnote 7 above, which deals with difficulties of the special theory of relativity and precedes the talk at which Feigl was present. Cf. also the letters to M. Besso and K. Seelig as quoted in G. Holton, 'Influences on Einstein's Early Work', *Organon*, No. 3, 1966, p. 242, and K. Seelig, *Albert Einstein*, Zurich, 1960, p. 271. In 1952 Born wrote to Einstein (*Born–Einstein Letters*, New York, 1971, p. 190, dealing with Freundlich's analysis of the bending of light near the sun and the red shift): 'It really looks as if your formula is not quite correct. It looks even worse in the case of the red shift [the crucial case referred to by Feigl and Popper]; this is much smaller than the theoretical value towards the centre of the sun's disk, and much larger at the edges. . . . Could this be a hint of non-linearity?' Einstein (letter of 12 May 1952, op. cit., p. 192) replied. 'Freundlich . . . does not move me in the slightest. Even if the deflection of light, the perihelial movement or line shift were unknown, the gravitation equations would still be convincing because they avoid the inertial system (the phantom which affects everything but is not itself affected). *It is really strange that human beings are normally deaf to the strongest arguments while they are always inclined to overestimate measuring accuracies*' (my italics). How is this conflict (between Feigl's testimony and Einstein's writings) to be explained? It cannot be explained by a *change* in Einstein's attitude. His disrespectful attitude towards observation and experiment was there from the very beginning, as we have seen. It might be explained either by a mistake

The second case, the case of qualitative failures, is less familiar, but of much greater interest. In this case a theory is inconsistent not with a recondite fact, that can be unearthed with the help of complex equipment and is known to experts only, but with circumstances which are easily noticed and which are familiar to everyone.

The first and, to my mind, the most important example of an inconsistency of this kind is Parmenides' theory of the unchanging and homogeneous One. This theory illustrates a desire that has propelled the Western sciences from their inception up to the present time – the desire to find a unity behind the many events that surround us. Today the unity sought is a *theory* rich enough to produce all the accepted facts and laws; at the time of Parmenides the unity sought was a *substance*. Thales had proposed water,[13] Heraclitus fire, Anaximander a substance which he called the *apeiron* and which could produce all four elements without being identical with a single one of them. Parmenides gave what seems to be an obvious and rather trivial answer: the substance that underlies everything that is is *Being*. But this trivial answer had surprising consequences. For example, we can assert that (first principle) *Being* is and that (second principle) *Not Being is not*. Now consider change and assume it to be fundamental. Then change can only go from Being to Not Being. But according to the second principle Not Being is not, which means that there is no fundamental change. Next consider difference and assume it to be fundamental. Then the difference can only be between Being and Not Being. But (second principle) Not Being is not and therefore there exist no differences in Being – it is a single, unchanging, continuous block. Parmenides knew of course that people, himself included, perceive and accept change and difference; but as his argument had shown that the perceived processes could not be fundamental he had

---

on Feigl's part, or else as another instance of Einstein's 'opportunism' – cf. text to footnote 6 of the Introduction.

On the last page (p. 91) of his *Über die Spezielle und allgemeine Relativitätstheorie*, Brunswick, 1922, Einstein writes: 'If the red shift of the spectral lines caused by the gravitational potential did not exist, then the general theory of relativity would be untenable.' Does this conflict with Einstein's cavalier attitude towards observation as described above? It does not. The passage speaks of *the red shift* not of *observations of it*.

13. The following account is highly speculative. Details in Vols 1 and 2 of W.K.C. Guthrie, *A History of Greek Philosophy*, Cambridge, 1962 and 1965, as well as in Chapters 1, 2 and 3 of my *Farewell to Reason*.

to regard them as merely apparent, or deceptive. This is indeed what he said – thus anticipating all those scientists who contrasted the 'real' world of science with the everyday world of qualities and emotions, declared the latter to be 'mere appearance' and tried to base their arguments on 'objective' experiments and mathematics exclusively. He also anticipated a popular interpretation of the theory of relativity which sees all events and transitions as already prearranged in a four-dimensional continuum, the only change being the (deceptive) journey of consciousness along its world line.[14] Be that as it may, he was the first to propose a conservation law (*Being is*), to draw a boundary line between reality and appearance (and thus to create what later thinkers called a 'theory of knowledge') and to give a more satisfactory foundation for continuity than did 19th- and 20th-century mathematicians who had to invoke 'intuition'. Using Parmenides' arguments Aristotle constructed a theory of space and motion that anticipated some very deep-lying properties of quantum mechanics and evaded the difficulties of the more customary (and less sophisticated) interpretation of a continuum as consisting of indivisible elements.[15] Parmenides' theory clashes with most modern methodological principles – but this is no reason to disregard it.

A more specific example of a theory with qualitative defects is Newton's theory of colours. According to this theory, light consists of rays of different refrangibility which can be separated, reunited, refracted, but which are never changed in their internal constitution, and which have a very small lateral extension in space. Considering that the surface of mirrors is much rougher than the lateral extension of the rays, the ray theory is found to be inconsistent with the existence of mirror images (as is admitted by Newton himself): if light consists of rays, then a mirror

---

14. A vivid description of the Parmenidean flavour of the theory of relativity is given by H. Weyl, *Philosophy of Mathematics and Natural Science*, Princeton, 1949, p. 116. Einstein himself wrote: 'For us who are convinced physicists the distinction between past, present and future has no other meaning than that of an illusion, though a tenacious one.' *Correspondence avec Michele Besso*, Paris, 1979, p. 312. See also p. 292. In a word: the events of a human life are 'illusions, though tenacious ones'.

15. For Aristotle see the essay quoted in Chapter 4, fn 3. Modern attempts to get continuity out of collections of indivisible elements are reported in A. Gruenbaum, 'A Consistent Conception of the Extended Linear Continuum as an Aggregate of Unextended Elements', *Philosophy of Science*, No. 19, 1952, pp. 283ff. See also W. Salmon (ed.), *Zeno's Paradoxes*, New York, 1970.

should behave like a rough surface, i.e. it should look to us like a wall. Newton retained his theory, eliminating the difficulty with the help of an *ad hoc* hypothesis: 'the reflection of a ray is effected, not by a single point of the reflecting body, but by some power of the body which is evenly diffused all over its surface'.[16]

In Newton's case the qualitative discrepancy between theory and fact was removed by an *ad hoc* hypothesis. In other cases not even this very flimsy manoeuvre is used: one retains the theory *and tries to forget* its shortcomings. An example is the attitude towards Kepler's rule according to which an object viewed through a lens is perceived at the point at which the rays travelling from the lens towards the eye intersect.[17]

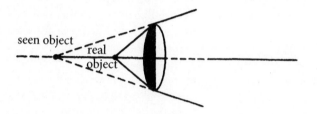

The rule implies that an object situated at the focus will be seen infinitely far away.

'But on the contrary,' writes Barrow, Newton's teacher and predecessor at Cambridge, commenting on this prediction,[18] 'we are assured by experience that [a point situated close to the focus] appears variously distant, according to the different situations of the eye. . . . And it does almost never seem further off than it would be if it were beheld with the naked eye; but, on the contrary, it does sometimes appear much

16. Sir Isaac Newton, *Optics*, Book 2, part 3, proposition 8, New York, 1952, p. 266. For a discussion of this aspect of Newton's method see my essay, 'Classical Empiricism', *Philosophical Papers*, Vol. 2, Chapter 2.

17. Johannes Kepler, *Ad Vitellionem Paralipomena, Johannes Kepler, Gesammelte Werke*, Vol. 2, Munich, 1939, p. 72. For a detailed discussion of Kepler's rule and its influence see Vasco Ronchi, *Optics: The Science of Vision*, New York, 1957, Chapters 43ff. See also Chapters 9–11 below.

18. *Lectiones XVIII Cantabrigiae in Scholio publicis habitae in quibus Opticorum Phenomenon genuinae Rationes investigantur ac exponentur*, London, 1669, p. 125. The passage is used by Berkeley in his attack on the traditional, 'objectivistic' optics (*An Essay Towards a New Theory of Vision*, Works, Vol. 1, ed. Frazer, London, 1901, pp. 137ff).

nearer. . . . All which does seem repugnant to our principles.' 'But for me,' Barrow continues, 'neither this nor any other difficulty shall have so great an influence on me, as to make me renounce that which I know to be manifestly agreeable to reason.'

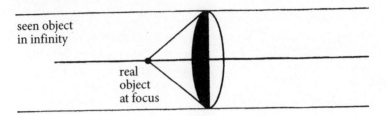

Barrow *mentions* the qualitative difficulties, adding that he will not abandon the theory. This is not the usual procedure. The usual procedure is to forget the difficulties, never to talk about them, and to proceed as if the theory were without fault. This attitude is very common today.

Thus the classical electrodynamics of Maxwell and Lorentz implies that the motion of a free particle is self-accelerated. Considering the self-energy of the electron one obtains divergent expressions for point-charges while charges of finite extension can be made to agree with relativity only by adding untestable stresses and pressures inside the electron.[19] The problem reappears in the quantum theory, though it is here partially covered up by 'renormalization'. This procedure consists in crossing out the results of certain calculations and replacing them by a description of what is actually observed. Thus one admits, implicitly, that the theory is in trouble while formulating it in a manner suggesting that a new principle has been discovered.[20] Small wonder when philosophically

19. See W. Heitler, *The Quantum Theory of Radiation*, Oxford, 1954, p. 31.

20. Renormalization has in the meantime become the basis of quantum field theory and has led to predictions of surprising accuracy (report with literature in A. Pais, *Inward Bound*, Oxford, 1986). This shows that a point of view which, looked at from afar, appears to be hopelessly wrong may contain excellent ingredients and that its excellence may remain unrevealed to those guided by strict methodological rules. Always remember that my examples do not criticize science; they criticize those who want to subject it to their simpleminded rules by showing the disasters such rules would create. Each of the examples of footnotes 3–17 can be used as a basis for case studies of the kind to be carried out in Chapters 6–12 (Galileo and the Copernican Revolution). This shows that the case of Galileo is not 'an exception characterizing the beginning of the so-called scientific revolution' (G.

unsophisticated authors get the impression that 'all evidence points with merciless definiteness in the ... direction ... [that] all the processes involving ... unknown interactions conform to the fundamental quantum law'.[21]

A striking example of qualitative failure is the status of classical mechanics and electrodynamics after Boltzmann's equipartition theorem. According to this theorem energy is equally distributed over all degrees of freedom of a (mechanical or electrodynamic) system. Both atoms (which had to be elastic to rebound from the walls of a container and from each other) and the electromagnetic field had infinitely many degrees of freedom, which meant that solids and the electromagnetic fields should have acted as insatiable sinks of energy. Yet '[a]s so often in the history of science, the conflict between simple and generally known facts and current theoretical ideas was recognized only slowly'.[22]

Another example of modern physics is quite instructive, for it might have led to an entirely different development of our knowledge concerning the microcosm. Ehrenfest proved a theorem according to which the classical electron theory of Lorentz taken in conjunction with the equipartition theorem excludes induced magnetism.[23] The reasoning is exceedingly simple; according to the equipartition theorem, the probability of a given motion is proportional to $\exp(-U/RT)$, where U is the energy of the motion. Now the rate of work of an electron moving in a constant magnetic field B is, according to Lorentz, $W=Q(E+V\times B)$. V, where Q is the charge of the moving particle, V its velocity and E the electric field. This magnitude reduces to QEV, which means that the

---

Radnitzky, 'Theorienpluralismus Theorienmonismus', in Diemer Meisenheim (ed.), *Der Methoden- und Theorienpluralismus in den Wissenschaften*, 1971, p. 164) but is *typical* of scientific change at all times.

21. Rosenfeld in *Observation and Interpretation*, London, 1957, p. 44.

22. K. Gottfried, V.F. Weisskopf, *Concepts of Particle Physics*, Vol. 1, Oxford and New York, 1984, p. 6.

23. The difficulty was realized by Bohr in his doctoral thesis; see Niels Bohr, *Collected Works*, Vol. 1, Amsterdam, 1972, pp. 158, 381. He pointed out that the velocity changes due to the changes in the external field would equalize after the field was established, so that no magnetic effects could arise. See also Heilbron and T.S. Kuhn, 'The Genesis of the Bohr Atom', *Historical Studies in the Physical Sciences*, No. 1, 1969, p. 221. The argument in the text is taken from *The Feynman Lectures*, Vol. 2, California and London, 1965, Chapter 34.6. For a somewhat clearer account cf. R. Becker, *Theorie der Elektrizität*, Leipzig, 1949, p. 132.

energy and, therefore, the probability remains unaffected by a magnetic field. (Given the proper context, this result strongly supports the ideas and experimental findings of the late Felix Ehrenhaft.)

Occasionally it is impossible to survey all the interesting consequences, and thus to discover the absurd results of a theory. This may be due to a deficiency in the existing mathematical methods; it may also be due to the ignorance of those who defend the theory. Under such circumstances, the most common procedure is to use an older theory up to a certain point (which is often quite arbitrary) and to add the new theory for calculating refinements. Seen from a methodological point of view the procedure is a veritable nightmare. Let us explain it using the relativistic calculation of the path of Mercury as an example.

The perihelion of Mercury moves along at a rate of about 5600" per century. Of this value, 5026" are geometric, having to do with the movement of the reference system, while 531" are dynamical, due to perturbations in the solar system. Of these perturbations all but the famous 43" are accounted for by classical mechanics. This is how the situation is usually explained.

The explanation shows that the premise from which we derive the 43" is not the general theory of relativity plus suitable initial conditions. The premise contains classical physics *in addition* to whatever relativistic assumptions are being made. Furthermore, the relativistic calculation, the so-called 'Schwarzschild solution', does not deal with the planetary system as it exists in the real world (i.e. our own asymmetric galaxy); it deals with the entirely fictional case of a central symmetrical universe containing a singularity in the middle and nothing else. What are the reasons for employing such an odd conjunction of premises?

The reason, according to the customary reply, is that we are dealing with approximations. The formulae of classical physics do not appear because relativity is incomplete. Nor is the centrally symmetric case used because relativity does not offer anything better. Both schemata flow from the general theory under the special circumstances realized in our planetary system *provided* we omit magnitudes too small to be considered. Hence, we are using the theory of relativity throughout, and we are using it in an adequate manner.

Now in the present case, making the required approximations would mean calculating the full n-body problem relativistically (including long-term resonances between different planetary orbits), omitting magnitudes smaller than the precision of observation reached, and showing that the theory thus curtailed coincides with classical celestial mechanics as corrected by Schwarzschild. This procedure has not been used by anyone simply because the relativistic n-body problem has as yet withstood solution. When the argument started, there were not even approximate solutions for important problems such as, for example, the problem of stability (one of the first great stumbling blocks for Newton's theory). The classical part of the explanans, therefore, did not occur just for convenience, *it was absolutely necessary*. And the approximations made were not a result of relativistic calculations, they were introduced in order to make relativity fit the case. One may properly call them *ad hoc approximations*.[24]

Ad hoc approximations abound in modern mathematical physics. They play a very important part in the quantum theory of fields and they are an essential ingredient of the correspondence principle. At the moment we are not concerned with the reasons for this fact, we are only concerned with its consequences: *ad hoc* approximations conceal, and even eliminate, qualitative difficulties. They create a false impression of the excellence of our science. It follows that a philosopher who wants to study the adequacy of science as a picture of the world, or who wants to build up a realistic scientific methodology, must look at modern science with special care. In most cases modern science is more opaque, and more deceptive, than its 16th- and 17th-century ancestors have ever been.

As a final example of qualitative difficulties I mention again the heliocentric theory at the time of Galileo. I shall soon have occasion to show that this theory was inadequate both qualitatively and quantitatively, and that it was also philosophically absurd.

---

24. Today the so-called parametrized post-Newtonian formalism satisfies most of the desiderata outlined in the text (details in Will, *Theory*). My point is that this was a later achievement whose absence did not prevent scientists from arguing, *and arguing well*, about the new ideas. Theories are not only used as premises for derivations; they are even more frequently used as a general background for novel guesses whose formal relation to the basic assumptions is difficult to ascertain. 'I must . . . confess', writes Descartes in his *Discourse on Method* (Library of Liberal Arts, 1965, p. 52), 'that the power of nature is so ample and so vast, and these principles [the theoretical principles he had developed for his mechanical universe] so simple and so general, that I almost never notice any particular effect such that I do not see right away that it can [be made to conform to these principles] in many different ways; and my greatest difficulty is usually to discover in which of these ways the effect is derived'. Modern theoretical physicists find themselves in exactly the same situation.

To sum up this brief and very incomplete list: wherever we look, whenever we have a little patience and select our evidence in an unprejudiced manner, we find that theories fail adequately to reproduce certain *quantitative results*, and that they are *qualitatively incompetent* to a surprising degree. Science gives us theories of great beauty and sophistication. Modern science has developed mathematical structures which exceed anything that has existed so far in coherence, generality and empirical success. But in order to achieve this miracle all the existing troubles had to be pushed into the *relation* between theory and fact,[25] and had to be concealed, by *ad hoc* hypotheses, *ad hoc* approximations and other procedures.

This being the case, what shall we make of the methodological demand that a theory must be judged by experience and must be rejected if it contradicts accepted basic statements? What attitude shall we adopt towards the various theories of confirmation and corroboration, which all rest on the assumption that theories can be made to agree with the known facts, and which use the amount of agreement reached as a principle of evaluation? This demand, these theories, are now all seen to be quite useless. They are as useless as a medicine that heals a patient only if he is bacteria-free. In practice they are never obeyed by anyone. Methodologists may point to the importance of falsifications – but they

---

25. Von Neumann's work in quantum mechanics is an especially instructive example of this procedure. In order to arrive at a satisfactory proof of the expansion theorem in Hilbert Space, von Neumann replaced the quasi-intuitive notions of Dirac (and Bohr) by more complex notions of his own. The theoretical relations between the new notions are accessible to a more rigorous treatment than the theoretical relations between the notions that preceded them ('more rigorous' from the point of view of von Neumann and his followers). It is different with their relation to experimental procedures. No measuring instruments can be specified for the great majority of observables (Wigner, *American Journal of Physics*, Vol. 31, 1963, p. 14), and where specification is possible it becomes necessary to modify well-known and unrefuted laws in an arbitrary way or else to admit that some quite ordinary problems of quantum mechanics, such as the scattering problem, do not have a solution (J.M. Cook, *Journal of Mathematical Physics*, Vol. 36, 1957). Thus the theory becomes a veritable monster of rigour and precision while its relation to experience is more obscure than ever. It is interesting to see that similar developments occur in 'primitive thought'. 'The most striking features of Nupe sand divining', writes S.F. Nader in *Nupe Religion*, 1954, p. 63, 'is the contrast between its pretentious theoretical framework and its primitive and slipshod application in practice.' It does not need a science to produce Neumannian nightmares.

blithely use falsified theories. They may sermonize how important it is to consider all the relevant evidence, and never mention those big and drastic facts which show that the theories they admire and accept may be as badly off as the older theories which they reject. In *practice* they slavishly repeat the most recent pronouncements of the top dogs in physics, though in doing so they must violate some very basic rules of their trade. Is it possible to proceed in a more reasonable manner? Let us see![26]

According to Hume, theories cannot be *derived from* facts. The demand to admit only those theories which follow from facts leaves us without any theory. Hence, science *as we know it* can exist only if we drop the demand and revise our methodology.

According to our present results, hardly any theory is *consistent with* the facts. The demand to admit only those theories which are consistent with the available and accepted facts again leaves us without any theory. (I repeat: *without any theory*, for there is not a single theory that is not in some trouble or other.) Hence, a science as we know it can exist only if we drop this demand also and again revise our methodology, *now admitting counterinduction in addition to admitting unsupported hypotheses*. The right method must not contain any rules that make us choose between theories *on the basis of falsification*. Rather, its rules must enable us to choose between theories which we have already tested *and which are falsified*.

To proceed further. Not only are facts and theories in constant disharmony, they are never as neatly separated as everyone makes them out to be. Methodological rules speak of 'theories', 'observations' and 'experimental results' as if these were well-defined objects whose properties are easy to evaluate and which are understood in the same way by all scientists.

However, the material which a scientist *actually* has at his disposal, his laws, his experimental results, his mathematical techniques, his epistemological prejudices, his attitude towards the absurd consequences of the theories which he accepts, is indeterminate in many ways,

---

26. The existence of qualitative difficulties, or 'pockets of resistance' (St Augustine, *Contra Julianum*, V, xiv, 51 – *Migne*, Vol. 44), was used by the Church fathers to defuse objections which the science of their time had raised against parts of the Christian faith, such as the doctrine of the corporeal resurrection.

ambiguous, *and never fully separated from the historical background*. It is contaminated by principles which he does not know and which, if known, would be extremely hard to test. Questionable views on cognition, such as the view that our senses, used in normal circumstances, give reliable information about the world, may invade the observation language itself, constituting the observational terms as well as the distinction between veridical and illusory appearance. As a result, observation languages may become tied to older layers of speculation which affect, in this roundabout fashion, even the most progressive methodology. (Example: the absolute space-time frame of classical physics which was codified and consecrated by Kant.) The sensory impression, however simple, contains a component that expresses the physiological reaction of the perceiving organism and has no objective correlate. This 'subjective' component often merges with the rest, and forms an unstructured whole which must be subdivided from the outside with the help of counterinductive procedures. (An example is the appearance of a fixed star to the naked eye, which contains the effects of irradiation diffraction, diffusion, restricted by the lateral inhibition of adjacent elements of the retina and is further modified in the brain.) Finally, there are the auxiliary premises which are needed for the derivation of testable conclusions, and which occasionally form entire *auxiliary sciences*.

Consider the case of the Copernican hypothesis, whose invention, defence, and partial vindication runs counter to almost every methodological rule one might care to think of today. The auxiliary sciences here contained laws describing the properties and the influence of the terrestrial atmosphere (meteorology); optical laws dealing with the structure of the eye and of telescopes, and with the behaviour of light; and dynamical laws describing motion in moving systems. Most importantly, however, the auxiliary sciences contained a theory of cognition that postulated a certain simple relation between perceptions and physical objects. Not all auxiliary disciplines were available in explicit form. Many of them merged with the observation language, and led to the situation described at the beginning of the preceding paragraph.

Consideration of all these circumstances, of observation terms, sensory core, auxiliary sciences, background speculation, suggest that a theory may be inconsistent with the evidence, not because it is incorrect, *but because the evidence is contaminated*. The theory is threatened because the evidence either contains unanalysed sensations which only partly correspond to external processes, or because it is presented in terms of antiquated views, or because it is evaluated with the help of backward auxiliary subjects. The Copernican theory was in trouble for *all* these reasons.

It is this *historico-physiological character of the evidence*, the fact that it does not merely describe some objective state of affairs *but also expresses subjective, mythical, and long-forgotten views* concerning this state of affairs, that forces us to take a fresh look at methodology. It shows that it would be extremely imprudent to let the evidence judge our theories directly and without any further ado. A straightforward and unqualified judgement of theories by 'facts' is bound to eliminate ideas *simply because they do not fit into the framework of some older cosmology.* Taking experimental results and observations for granted and putting the burden of proof on the theory means taking the observational ideology for granted without having ever examined it. (Note that the experimental results are supposed to have been obtained with the greatest possible care. Hence 'taking observations, etc., for granted' means 'taking them for granted *after* the most careful examination of their reliability': for even the most careful examination of an observation statement does not interfere with the concepts in which it is expressed, or with the structure of the sensory image.)

Now – how can we possibly examine something we use all the time and presuppose in every statement? How can we criticize the terms in which we habitually express our observations? Let us see!

The first step in our criticism of commonly-used concepts is to create a measure of criticism, something with which these concepts can be *compared*. Of course, we shall later want to know a little more about the measuring-stick itself; for example, we shall want to know whether it is better than, or perhaps not as good as, the material examined. But in order for *this* examination to start there must be a measuring-stick in the first place. Therefore, the first step in our criticism of customary concepts and customary reactions is to step outside the circle and either to invent a new conceptual system, for example a new theory, that clashes with the most carefully established observational results and confounds the most plausible theoretical principles, or to import such a system from outside science, from religion, from mythology, from the ideas of incompetents,[27] or the ramblings of madmen. This step is, again, counterinductive. Counterinduction is thus both a *fact* – science could not exist without it – and a legitimate and much needed *move* in the game of science.

---

27. It is interesting to see that Philolaos, who disregarded the evidence of the senses and set the earth in motion, was 'an unmathematical confusionist. It was the confusionist who found the courage lacking in many great observers and mathematically well-informed scientists to disregard the immediate evidence of the senses in order to remain in agreement with principles he firmly believed.' K. von Fritz, *Grundprobleme der Geschichte der antiken Wissenschaft*, Berlin-New York, 1971, p. 165. 'It is therefore not surprising that the next step on this path was due to a man whose writings, as far as we know them, show him as a talented stylist and popularizer with occasionally interesting ideas of his own rather than as a profound thinker or exact scientist,' ibid., p. 184. Confusionists and superficial intellectuals *move ahead* while the 'deep' thinkers *descend* into the darker regions of the status quo or, to express it in a different way, they remain stuck in the mud.

6

As an example of such an attempt I examine the tower argument *which the Aristotelians used to refute the motion of the earth. The argument involves* natural interpretations *– ideas so closely connected with observations that it needs a special effort to realize their existence and to determine their content. Galileo identifies the natural interpretations which are inconsistent with Copernicus and replaces them by others.*

It seems to me that [Galileo] suffers greatly from continual digressions, and that he does not stop to explain all that is relevant at each point; which shows that he has not examined them in order, and that he has merely sought reasons for particular effects, without having considered . . . first causes . . .; and thus that he has built without a foundation.

Descartes

I am (indeed) unwilling to compress philosophical doctrines into the most narrow kind of space and to adopt that stiff, concise and graceless manner, that manner bare of any adornment which pure geometricians call their own, not uttering a single word that has not been given to them by strict necessity . . . I do not regard it as a fault to talk about many diverse things, even in those treatises which have only a single topic . . . for I believe that what gives grandeur, nobility, and excellence to our deeds and inventions does not lie in what is necessary – though the absence of it would be a great mistake – but in what is not. . . .

Galileo

But where common sense believes that rationalizing sophists have the intention of shaking the very fundament of the commonweal, then it would seem to be not only reasonable, but permissible, and even praiseworthy to aid the good cause with sham reasons rather than leaving the advantage to the . . . opponent.

Kant[1]

---

1. The three quotations are: Descartes, letter to Mersenne of 11 October 1638, *Oeuvres*, 11, p. 380. Galileo, letter to Leopold of Toscana of 1640, usually quoted under the title *Sul*

As a concrete illustration and as a basis for further discussion, I shall now briefly describe the manner in which Galileo defused an important argument against the idea of the motion of the earth. I say 'defused', and not 'refuted', because we are dealing with a changing conceptual system as well as with certain attempts at concealment.

According to the argument which convinced Tycho, and which is used against the motion of the earth in Galileo's own *Trattato della sfera*, observation shows that 'heavy bodies . . . falling down from on high, go by a straight and vertical line to the surface of the earth. This is considered an irrefutable argument for the earth being motionless. For, if it made the diurnal rotation, a tower from whose top a rock was let fall, being carried by the whirling of the earth, would travel many hundreds of yards to the east in the time the rock would consume in its fall, and the rock ought to strike the earth that distance away from the base of the tower.'[2]

In considering the argument, Galileo at once admits the correctness of the sensory content of the observation made, viz. that 'heavy bodies . . . falling from a height, go perpendicularly to the surface of the earth.'[3] Considering an author (Chiaramonti) who sets out to convert Copernicus by repeatedly mentioning this fact, he says: 'I wish that this author would not put himself to

---

Candor Lunare, Opere, Favoro, VIII, p. 491. For a detailed discussion of Galileo's style and its connection with his natural philosophy see L. Olschki, *Galileo und seine Zeit: Geschichte der neusprachlichen wissenschaftlichen Literatur*, Vol. III, Halle, 1927, reprinted Vaduz, 1965. The letter to Leopold is quoted and discussed on pp. 455ff.

Descartes' letter is discussed by Salmon as an example of the issue between rationalism and empiricism in 'The Foundations of Scientific Inference', *Mind and Cosmos*, ed. Colodny, Pittsburgh, 1966, p. 136. It should rather be regarded as an example of the issue between dogmatic methodologies and opportunistic methodologies, bearing in mind that empiricism can be as strict and unyielding as the most rigorous types of rationalism.

The Kant quotation is from the *Critique of Pure Reason*, B777, p. 8ff (the quotation was brought to my attention by Professor Stanley Rosen's work on Plato's *Symposium*). Kant continues: 'However, I would think that there is nothing that goes less well together with the intention of asserting a good cause than subterfuge, conceit, and deception. *If* one could take only this much for granted, then the battle of speculative reason . . . would have been concluded long ago, or would soon come to an end. Thus the purity of a cause often stands in the inverse proportion to its truth. . . .' One should also note that Kant explains the rise of *civilization* on the basis of disingenuous moves which 'have the function to raise mankind above its crude past', ibid., 776, p. 14f. Similar ideas occur in his account of world history.

2. *Dialogue*, op. cit., p. 126.

3. Ibid., p. 125.

such trouble trying to have us understand from our senses that this motion of falling bodies is simple straight motion and no other kind, nor get angry and complain because such a clear, obvious, and manifest thing should be called into question. For in this way he hints at believing that to those who say such motion is not straight at all, but rather circular, it seems they see the stone move visibly in an arc, since he calls upon their senses rather than their reason to clarify the effect. This is not the case, Simplicio; for just as I . . . have never seen nor ever expect to see, the rock fall any way but perpendicularly, just so do I believe that it appears to the eyes of everyone else. It is, therefore, better to put aside the appearance, on which we all agree, and to use the power of reason either to confirm its reality or to reveal its fallacy.'[4] The correctness of the observation is not in question. What is in question is its 'reality' or 'fallacy'. What is meant by this expression?

The question is answered by an example that occurs in Galileo's next paragraph, 'from which . . . one may learn how easily anyone may be deceived by simple appearance, or let us say by the impressions of one's senses. This event is the appearance to those who travel along a street by night of being followed by the moon, with steps equal to theirs, when they see it go gliding along the eaves of the roofs. There it looks to them just as would a cat really running along the tiles and putting them behind it; an appearance which, if reason did not intervene, would only too obviously deceive the senses.'

In this example, we are asked to start with a sensory impression and to consider a statement that is forcefully suggested by it. (The suggestion is so strong that it has led to entire systems of belief and to rituals, as becomes clear from a closer study of the lunar aspects of witchcraft and of other cosmological hypotheses.) Now 'reason intervenes'; the statement suggested by the impression is examined, and one considers other statements in its place. The nature of the impression is not changed a bit by this activity. (This is only approximately true; but we can omit for our present purpose the complications arising from an interaction of impression and proposition.) But it enters new observation statements and plays new, better or worse, parts in our knowledge. What are the reasons and the methods which regulate such an exchange?

To start with, we must become clear about the nature of the total phenomenon: appearance plus statement. There are not two acts – one, noticing a phenomenon; the other, expressing it with the help of the appropriate statement – *but only one*, viz. saying in a certain

---

4. Ibid., p. 256.

observational situation, 'the moon is following me', or, 'the stone is falling straight down'. We may, of course, abstractly subdivide this process into parts, and we may also try to create a situation where statement and phenomenon seem to be psychologically apart and waiting to be related. (This is rather difficult to achieve and is perhaps entirely impossible.) But under normal circumstances such a division does not occur; describing a familiar situation is, for the speaker, an event in which statement and phenomenon are firmly glued together.

This unity is the result of a process of learning that starts in one's childhood. From our very early days we learn to react to situations with the appropriate responses, linguistic or otherwise. The teaching procedures both *shape* the 'appearance', or 'phenomenon', and establish a firm *connection* with words, so that finally the phenomena seem to speak for themselves without outside help or extraneous knowledge. They *are* what the associated statements assert them to be. The language they 'speak' is, of course, influenced by the beliefs of earlier generations which have been held for so long that they no longer appear as separate principles, but enter the terms of everyday discourse, and, after the prescribed training, seem to emerge from the things themselves.

At this point we may want to compare, in our imagination and quite abstractly, the results of the teaching of different languages incorporating different ideologies. We may even want consciously to change some of these ideologies and adapt them to more 'modern' points of view. It is very difficult to say how this will alter our situation, *unless* we make the further assumption that the quality and structure of sensations (perceptions), or at least the quality and structure of those sensations which enter the body of science, is independent of their linguistic expression. I am very doubtful about even the approximate validity of this assumption, which can be refuted by simple examples, and I am sure that we are depriving ourselves of new and surprising discoveries as long as we remain within the limits defined by it. Yet, I shall for the moment, remain within these limits.

Making the additional simplifying assumption, we can now distinguish between sensations and those 'mental operations which follow so closely upon the senses',[5] and which are so firmly connected with their reactions that a separation is difficult to achieve. Considering the origin and the effect of such operations, I shall call them *natural interpretations*.

---

5. Francis Bacon, *Novum Organum*, Introduction.

In the history of thought, natural interpretations have been regarded either as *a priori presuppositions* of science, or else as *prejudices* which must be removed before any serious examination can begin. The first view is that of Kant, and, in a very different manner and on the basis of very different talents, that of some contemporary linguistic philosophers. The second view is due to Bacon (who had predecessors, however, such as the Greek sceptics).

Galileo is one of those rare thinkers who wants neither forever to *retain* natural interpretations nor altogether to *eliminate* them. Wholesale judgements of this kind are quite alien to his way of thinking. He insists upon a *critical discussion* to decide which natural interpretations can be kept and which must be replaced. This is not always clear from his writings. Quite the contrary. The methods of reminiscence, to which he appeals so freely, are designed to create the impression that nothing has changed and that we continue expressing our observations in old and familiar ways. Yet his attitude is relatively easy to ascertain: natural interpretations are *necessary*. The senses alone, without the help of reason, cannot give us a true account of nature. What is needed for arriving at such a true account are 'the . . . senses, *accompanied by reasoning*'.[6] Moreover, in the arguments dealing with the motion of the earth, it is this reasoning, it is the connotation of the observation terms and *not* the message of the senses or the appearance, that causes trouble. 'It is, therefore, better to put aside the appearance, on which we all agree, and to use the power of reason either to confirm its reality or to reveal its fallacy.'[7] Confirming the reality or revealing the fallacy of appearances means, however, examining the validity of those natural interpretations which are so intimately connected with the appearances that we no longer regard them as separate assumptions. I now turn to the first natural interpretation implicit in the argument from falling stones.

According to the Copernican view as presupposed in the tower argument, the motion of a falling stone should be 'mixed straight-and-circular'.[8] By the 'motion of the stone' is meant not its motion relative to some visible mark in the visual field of the observer, or its observed motion, but rather its motion in the solar system or in (absolute) space, i.e. its *real motion*. The familiar facts appealed to in the argument present

---

6. *Dialogue*, op. cit., p. 255, my italics.
7. Ibid., p. 256.
8. Ibid., p. 248.

a different kind of motion, a simple vertical motion. This refutes the Copernican hypothesis only if the concept of motion that occurs in the observation statement is the same as the concept of motion that occurs in the Copernican prediction. The observation statement 'the stone is falling straight down' must, therefore, refer to a movement in (absolute) space. It must refer to a real motion.

Now, the force of an 'argument from observation' derives from the fact that the observation statements involved are firmly connected with appearances. There is no use appealing to observation if one does not know how to describe what one sees, or if one can offer one's description with hesitation only, as if one had just learned the language in which it is formulated. Producing an observation statement, then, consists of two very different psychological events: (1) a clear and unambiguous *sensation* and (2) a clear and unambiguous *connection* between this sensation and parts of a language. This is the way in which the sensation is made to speak. Do the sensations in the above argument speak the language of real motion?

They speak the language of real motion in the context of 17th-century everyday thought. At least, this is what Galileo tells us. He tells us that the everyday thinking of the time assumes the 'operative' character of *all* motion, or, to use well-known philosophical terms, it assumes *a naive realism with respect to motion*: except for occasional and unavoidable illusions, apparent motion is identical with real (absolute) motion. Of course, this distinction is not explicitly drawn. One does not first distinguish the apparent motion from the real motion and then connect the two by a correspondence rule. One rather describes, perceives, acts towards motion as if it were already the real thing. Nor does one proceed in this manner under all circumstances. It is admitted that objects may move which are not seen to move; and it is also admitted that certain motions are illusory (cf. the example of the moon mentioned earlier in this chapter). Apparent motion and real motion are not always identified. However, there are *paradigmatic cases* in which it is psychologically very difficult, if not plainly impossible, to admit deception. It is from these paradigmatic cases, and not from the exceptions, that naive realism with respect to motion derives its strength. These are also the situations in which we first learn our kinematic vocabulary. From our very childhood we learn to react to them with concepts which have naive realism built right into them, and which inextricably connect movement and the appearance of movement. The motion of the stone in the tower argument, or the alleged motion of the earth, is such a paradigmatic case. How could one possibly be unaware of the swift motion of a large bulk of matter such as the earth is supposed to be! How could one possibly be unaware of the fact that the falling stone traces a vastly extended trajectory through space! From the point of view of 17th-century thought and language, the argument is, therefore, impeccable and quite forceful. However, notice how *theories* ('operative character' of all motion; essential correctness of sense reports) which are not formulated explicitly, enter the debate in the guise of observable events. We realize again that such events are Trojan horses which must be watched most carefully. How is one supposed to proceed in such a sticky situation?

The argument from falling stones seems to refute the Copernican view. This may be due to an inherent disadvantage of Copernicanism; but it may also be due to the presence of natural interpretations which are in need of improvement. The first task, then, is to *discover* and to isolate these unexamined obstacles to progress.

It was Bacon's belief that natural interpretations could be discovered by a method of analysis that peels them off, one after another, until the sensory core of every observation is laid bare. This method has serious drawbacks. First, natural interpretations of the kind considered by Bacon are not just *added* to a previously existing field of sensations. They are instrumental in *constituting* the field, as Bacon says himself. Eliminate all natural interpretations, and you also eliminate the ability to think and to perceive. Second, disregarding this fundamental function of natural interpretations, it should be clear that a person who faces a perceptual field without a single natural interpretation at his disposal would be *completely disoriented*, he could not even *start* the business of science. The fact that we *do* start, even after some Baconian analysis, therefore shows that the analysis has stopped prematurely. It has stopped at precisely those natural interpretations of which we are not aware and without which we cannot proceed. It follows that the intention to start from scratch, after a complete removal of all natural interpretations, is self-defeating.

Furthermore, it is not possible even *partly* to unravel the cluster of natural interpretations. At first sight the task would seem to be simple enough. One takes observation statements, one after the other, and analyses their content. However, concepts that are hidden in observation statements are not likely to reveal themselves in the more abstract parts of language. If they do, it will still be difficult to nail them down; concepts, just like percepts, are ambiguous and dependent on background. Moreover, the content of a concept is determined also by the way in which it is related to perception. Yet, how can this way be discovered without circularity? Perceptions must be identified, and the identifying mechanism will contain some of the very same elements which govern the use of the concept to be investigated. We never penetrate this concept completely, for we always use part of it in the attempt to find its constituents. There is only one way to get out of this circle, and it consists in using an *external measure of comparison*, including new ways of relating concepts and percepts. Removed from the domain of natural discourse and from all those principles, habits, and attitudes which constitute its form of life, such an external measure will look strange indeed. This, however, is not an argument against its use. On the contrary, such an impression of strangeness reveals that natural interpretations are at work, and is a first step towards their discovery. Let us explain this situation with the help of the tower example.

The example is intended to show that the Copernican view is not in accordance with 'the facts'. Seen from the point of view of these 'facts', the idea of the motion of the earth is outlandish, absurd, and obviously false, to mention only some of the expressions which were frequently used at the time, and which are still heard whenever professional squares confront a new and counter-factual theory. This makes us suspect that the Copernican view is an external measuring rod of precisely the kind described above.

Let us therefore turn the argument around and use it as a *detecting device* that helps us to discover the natural interpretations which exclude the motion of the earth. Turning the argument around, we *first assert* the motion of the earth and *then inquire* what changes will remove the contradiction. Such an inquiry may take considerable time, and there is a good sense in which it is not finished even today. The contradiction may stay with us for decades or even centuries. Still, *it must be upheld* until we have finished our examination or else the examination, the attempt to discover the antediluvian components of our knowledge, cannot even start. This, we have seen, is one of the reasons one can give for *retaining*, and, perhaps, even for *inventing*, theories which are inconsistent with the facts. Ideological ingredients of our knowledge and, more especially, of our observations are discovered with the help of theories which are refuted by them. *They are discovered counterinductively.*

Let me repeat what has been asserted so far. Theories are tested, and possibly refuted, by facts. Facts contain ideological components, older views which have vanished from sight or were perhaps never formulated in an explicit manner. Such components are highly suspicious. First, because of their age and obscure origin: we do not know why and how they were introduced; secondly, because their very nature protects them, and always has protected them, from critical examination. In the event of a contradiction between a new and interesting theory and a collection of firmly established facts, the best procedure, therefore, is not to abandon the theory but to use it to discover the hidden principles responsible for the contradiction. Counterinduction is an essential part of such a process of discovery. (Excellent historical example: the arguments against motion and atomicity of Parmenides and Zeno. Diogenes of Sinope, the Cynic, took the simple course that would be taken by many contemporary scientists and all contemporary philosophers: he refuted the arguments by rising and walking up and down. The opposite course, recommended here, has led to much more interesting results, as is witnessed by the history of the case. One should not be too hard on Diogenes, however, for it is also reported that he beat up a pupil who was content with his refutation, exclaiming that he had given reasons which the pupil should not accept without additional reasons of his own.[9])

Having *discovered* a particular natural interpretation, how can we *examine* it and *test* it? Obviously, we cannot proceed in the usual way, i.e. derive predictions and compare them with 'results of observation'. These results are no longer available. The idea that the senses, employed under normal circumstances, produce correct reports of real events, for example reports of the real motion of physical bodies, has been removed from all observational statements. (Remember that this notion was found to be an essential part of the anti-Copernican argument.) But without it our sensory reactions cease to be relevant for tests. This conclusion was generalized by some older rationalists, who decided to build their science on reason only and ascribed to observation a quite insignificant auxiliary function. Galileo does not adopt this procedure.

If *one* natural interpretation causes trouble for an attractive view, and if its *elimination* removes the view from the domain of observation, then the

---

9. Hegel, *Vorlesungen über die Geschichte der Philosophie*, I, ed. C.L. Michelet, Berlin, 1840, p. 289.

only acceptable procedure is to use *other* interpretations and to see what happens. The interpretation which Galileo uses restores the senses to their position as instruments of exploration, *but only with respect to the reality of relative motion*. Motion 'among things which share it in common' is 'non-operative', that is, 'it remains insensible, imperceptible, and without any effect whatever'.[10] Galileo's first step, in his joint examination of the Copernican doctrine and of a familiar but hidden natural interpretation, consists therefore in *replacing the latter by a different interpretation*. In other words, *he introduces a new observation language*.

This is, of course, an entirely legitimate move. In general, the observation language which enters an argument has been in use for a long time and is quite familiar. Considering the structure of common idioms on the one hand, and of the Aristotelian philosophy on the other, neither this use nor this familiarity can be regarded as a test of the underlying principles. These principles, these natural interpretations, occur in every description. Extraordinary cases which might create difficulties are defused with the

---

10. *Dialogue*, op. cit., p. 171. Galileo's kinematic relativism is not consistent. In the passage quoted, he proposes the view (1) that shared motion has *no effect whatsoever*. 'Motion,' he says, 'in so far as it is and acts as motion, to that extent exists relatively to things that lack it; and among things which all share equally in any motion, it does not act and is as if it did not exist' (p. 116); 'Whatever motion comes to be attributed to the earth must necessarily remain imperceptible . . . so long as we look only at terrestrial objects' (p. 114); '. . . motion that is common to many moving things is idle and inconsequential to the relation of those movables among themselves . . .' (p. 116). On the other hand, (2) he also suggests that 'nothing . . . *moves in a straight line by nature*. The motion of all celestial objects is in a circle; ships, coaches, horses, birds, all move in a circle around the earth; the motions of the parts of animals are all circular; in sum – we are forced to assume that only *gravia deorsum* and *levia sursum* move apparently in a straight line; but even that is not certain as long as it has not been proven that the earth is at rest' (p. 19). Now, if (2) is adopted, then the loose parts of systems moving in a straight line will tend to describe circular paths, thus contradicting (1). It is this inconsistency which has prompted me to split Galileo's argument into two steps, one dealing with the relativity of motion (only relative motion *is noticed*), the other dealing with inertial laws (and only inertial motion *leaves the relation between the parts of a system unaffected* – assuming, of course, that neighbouring inertial motions are approximately parallel). For the two steps of the argument, see the next chapter. One must also realize that accepting relativity of motion for inertial paths means giving up the *impetus theory*, which provides an (inner) cause for motions and therefore assumes an absolute space in which this cause becomes manifest. This Galileo seems to have done by now, for his argument for the existence of 'boundless' or 'perpetual' motions which he outlines on pp. 147ff of the *Dialogue* appeals to motions which are neutral, i.e. neither natural nor forced, and which may therefore (?) be assumed to go on for ever.

help of 'adjustor words',[11] such as 'like' or 'analogous', which divert them so that the basic ontology remains unchallenged. A test is, however, urgently needed. It is especially needed in those cases where the principles seem to threaten a new theory. It is then quite reasonable to introduce alternative observation languages and to compare them both with the original idiom and with the theory under examination. Proceeding in this way, we must make sure that the comparison is *fair*. That is, we must not criticize an idiom that is supposed to function as an observation language because it is not yet well known and is, therefore, less strongly connected with our sensory reactions and less plausible than is another, more 'common' idiom. Superficial criticisms of this kind, which have been elevated into an entire 'philosophy', abound in discussions of the mind–body problem. Philosophers who want to introduce and to test new views thus find themselves faced not with *arguments*, which they could most likely answer, but with an impenetrable stone wall of well-entrenched *reactions*. This is not at all different from the attitude of people ignorant of foreign languages, who feel that a certain colour is much better described by 'red' than by 'rosso'. As opposed to such attempts at conversion by appeal to familiarity ('I *know* what pains are, and I also *know*, from introspection, that they have nothing whatever to do with material processes!'), we must emphasize that a comparative judgement of observation languages, e.g. materialistic observation languages, phenomenalistic observation languages, objective-idealistic observation languages, theological observation languages, etc., can start only *when all of them are spoken equally fluently*.

Let us now continue with our analysis of Galileo's reasoning.

---

11. J.L. Austin, *Sense and Sensibilia*, New York, 1964, p. 74. Adjustor words play an important role in the Aristotelian philosophy.

7

The new natural interpretations constitute a new and highly abstract observation language. They are introduced and concealed so that one fails to notice the change that has taken place (method of anamnesis). They contain the idea of the relativity of all motion and the law of circular inertia.

Galileo replaces one natural interpretation by a very different and as yet (1630) at least partly unnatural interpretation. How does he proceed? How does he manage to introduce absurd and counter-inductive assertions such as the assertion that the earth moves, and yet get them a just and attentive hearing? One anticipates that arguments will not suffice – an interesting and highly important limitation of rationalism – and Galileo's utterances are indeed arguments in appearance only. For Galileo uses *propaganda*. He uses *psychological tricks* in addition to whatever intellectual reasons he has to offer. These tricks are very successful: they lead him to victory. But they obscure the new attitude towards experience that is in the making, and postpone for centuries the possibility of a reasonable philosophy. They obscure the fact that the experience on which Galileo wants to base the Copernican view is nothing but the result of his own fertile imagination, that it has been *invented*. They obscure this fact by insinuating that the new results which emerge are known and conceded by all, and need only be called to our attention to appear as the most obvious expression of the truth.

Galileo 'reminds' us that there are situations in which the non-operative character of shared motion is just as evident and as firmly believed as the idea of the operative character of all motion is in other circumstances. (This latter idea is, therefore, not the only natural interpretation of motion.) The situations are: events in a boat, in a smoothly moving carriage, and in other systems that contain an observer and permit him to carry out some simple operations.

Sagredo: There has just occurred to me a certain fantasy which passed through my imagination one day while I was sailing to Aleppo, where I was going as consul for our country. . . . If the point of a pen had been on the ship during my whole voyage from Venice to Alexandretta and had

had the property of leaving visible marks of its whole trip, what trace –
what mark – what line would it have left?

Simplicio: It would have left a line extending from Venice to there; not
perfectly straight – or rather, not lying in the perfect arc of a circle – but
more or less fluctuating according as the vessel would now and again have
rocked. But this bending in some places a yard or two to the right or left,
up or down, in length of many hundreds of miles, would have made little
alteration in the whole extent of the line. These would scarcely be sensible,
and, without an error of any moment, it could be called part of a perfect
arc.

Sagredo: So that if the fluctuation of the waves were taken away and
the motion of the vessel were calm and tranquil, the true and precise
motion of that pen would have been an arc of a perfect circle. Now if I had
had that same pen continually in my hand, and had moved it only a little
sometimes this way or that, what alterations should I have brought into
the main extent of this line?

Simplicio: Less than that which would be given to a straight line a
thousand yards long which deviated from absolute straightness here and
there by a flea's eye.

Sagredo: Then if an artist had begun drawing with that pen on a sheet
of paper when he left the port and had continued doing so all the way to
Alexandretta, he would have been able to derive from the pen's motion
a whole narrative of many figures, completely traced and sketched in
thousands of directions, with landscapes, buildings, animals, and other
things. Yet the actual real essential movement marked by the pen point
would have been only a line; long, indeed, but very simple. But as to the
artist's own actions, these would have been conducted exactly the same as
if the ship had been standing still. The reason that of the pen's long motion
no trace would remain except the marks drawn upon the paper is that the
gross motion from Venice to Alexandretta was common to the paper, the
pen, and everything else in the ship. But the small motions back and forth,
to right and left, communicated by the artist's fingers to the pen but not to
the paper, and belonging to the former alone, could thereby leave a trace
on the paper which remained stationary to those motions.[1]

Or

---

1. *Dialogue*, op. cit., pp. 171ff.

Salviati: . . . Imagine yourself in a boat with your eyes fixed on a point of the sail yard. Do you think that because the boat is moving along briskly, you will have to move your eyes in order to keep your vision always on that point of the sail and follow its motion?

Simplicio: I am sure that I should not need to make any change at all; not just as to my vision, but if I had aimed a musket I should never have to move it a hairsbreadth to keep it aimed, no matter how the boat moved.

Salviati: And this comes about because the motion which the ship confers upon the sail yard, it confers also upon you and upon your eyes, so that you need not move them a bit in order to gaze at the top of the sail yard, which consequently appears motionless to you. (And the rays of vision go from the eye to the sail yard just as if a cord were tied between the two ends of the boat. Now a hundred cords are tied at different fixed points, each of which keeps its place whether the ship moves or remains still.)[2]

It is clear that these situations lead to a non-operative concept of motion even within common sense.

On the other hand, common sense, and I mean 17th-century Italian-artisan common sense, also contains the idea of the *operative* character of all motion. This latter idea arises when a limited object that does not contain too many parts moves in vast and stable surroundings; for example, when a camel trots through the desert, or when a stone descends from a tower.

Now Galileo urges us to 'remember' the conditions in which we assert the non-operative character of shared motion in this case also, and to subsume the second case under the first.

Thus, the first of the two paradigms of non-operative motion mentioned above is followed by the assertion that – 'It is likewise true that the earth being moved, the motion of the stone in descending is actually a long stretch of many hundred yards, or even many thousand; and had it been able to mark its course in motionless air or upon some

---

2. Ibid., pp. 249ff. That phenomena of *seen* motion depend on *relative* motion has been asserted by Euclid in his *Optics*, Theon red. par. 49ff. An old scholion of par. 50 uses the example of a boat leaving the harbour: Heiberg, VII, 283. The example is repeated by Copernicus in Book 1, Chapter VIII, of *De Revol.* It was a commonplace in mediaeval optics. Cf. Witelo, *Perspectiva*, iv par 138 (Basel, 1572, p. 180).

other surface, it would have left a very long slanting line. But that part of all this motion which is common to the rock, the tower, and ourselves remains insensible and as if it did not exist. There remains observable only that part in which neither the tower nor we are participants; in a word, that with which the stone, in falling, measures the tower.'[3]

And the second paradigm precedes the exhortation to 'transfer this argument to the whirling of the earth and to the rock placed on top of the tower, whose motion you cannot discern because, in common with the rock, you possess from the earth that motion which is required for following the tower; you do not need to move your eyes. Next, if you add to the rock a downward motion which is peculiar to it and not shared by you, and which is mixed with this circular motion, the circular portion of the motion which is common to the stone and the eye continues to be imperceptible. The straight motion alone is sensible, for to follow that you must move your eyes downwards.'[4]

This is strong persuasion indeed.

Yielding to this persuasion, we now *quite automatically* start confounding the conditions of the two cases and become relativists. This is the essence of Galileo's trickery! As a result, the clash between Copernicus and 'the conditions affecting ourselves and those in the air above us'[5] dissolves into thin air, and we finally realize 'that all terrestrial events from which it is ordinarily held that the earth stands still and the sun and the fixed stars are moving would necessarily appear just the same to us if the earth moved and the other stood still'.[6]

---

3. Ibid., pp. 172ff.
4. Ibid., p. 250.
5. Ptolemy, *Syntaxis*, i, 1, p. 7.
6. *Dialogue*, op. cit., p. 416: cf. the *Dialogues Concerning Two New Sciences*, transl. Henry Crew and Alfonso de Salvio, New York, 1958, p. 164: 'The same experiment which at first glance seemed to show one thing, when more carefully examined, assures us of the contrary.' Professor McMullin, in a critique of this way of seeing things, wants more 'logical and biographical justification' for my assertion that Galileo not only argued, but also cheated ('A Taxonomy of the Relation between History and Philosophy of Science', *Minnesota Studies*, Vol. 5, Minneapolis, 1971, p. 39), and he objects to the way in which I let Galileo introduce dynamical relativism. According to him, 'what Galileo argues is that since his opponent *already* interprets observations made in such a context [movements on boats] in a "relativistic" way, how can he consistently do otherwise in the case of observations made on the earth's surface?' (ibid., p. 40). This is indeed how Galileo argues. But he argues so against

Let us now look at the situation from a more abstract point of view. We start with two conceptual sub-systems of 'ordinary' thought (see table p. 67). One of them regards motion as an absolute process which always has effects, effects on our senses included. The description of this conceptual system given here may be somewhat idealized; but the arguments of Copernicus' opponents, which are quoted by Galileo himself and, according to him, are 'very plausible',[7] show that there was a widespread tendency to think in its terms, and that this tendency was a serious obstacle to the discussion of alternative ideas. Occasionally, one finds even more primitive ways of thinking, where concepts such as 'up' and 'down' are used absolutely. Examples are: the assertion 'that the earth is too heavy to climb up over the sun and then fall headlong back down again',[8] or the assertion that 'after a short time the mountains, sinking downward with the rotation of the terrestrial globe, would get into such a position that whereas a little earlier one would have had to climb steeply to their peaks, a few hours later one would have to stoop and descend in order to get there'.[9] Galileo, in his marginal notes, calls these 'utterly childish reasons [which] sufficed to keep imbeciles believing in the fixity of the earth',[10] and he thinks it unnecessary 'to bother about such men as those, *whose name is legion*, or to take notice of their fooleries'.[11] Yet it is clear that the absolute idea of motion was

---

an opponent who, according to him, 'feels a great repugnance towards recognizing this non-operative quality of motion among the things which share it in common' (*Dialogue*, op. cit., p. 171), who is convinced that a boat, apart from having relative motions, *has absolute positions and motions as well* (cf. Aristotle, Physics, 208b8ff), and who at any rate has developed the art of using different notions on different occasions without running into a contradiction. Now if *this* is the position to be attacked, then showing that an opponent has a relative idea of motion, or frequently uses the relative idea in his everyday affairs, is not at all 'proof of inconsistency in his own "paradigm"' (McMullin, op. cit., p. 40). It just reveals one part of that paradigm without touching the other. The argument turns into the desired proof only if the absolute notion is either suppressed or spirited away, or else identified with the relativistic notion – and this is what Galileo actually does, though surreptitiously, as I have tried to show.

7. *Dialogue*, op. cit., p. 328.
8. Ibid., p. 327
9. Ibid., p. 330.
10. Ibid., p. 327.
11. Ibid., p. 327, italics added.

'well-entrenched', and that the attempt to replace it was bound to encounter strong resistance.[12]

The second conceptual system is built around the relativity of motion, and is also well-entrenched in its own domain of application. Galileo aims at replacing the first system by the second in *all* cases, terrestrial as well as celestial. Naive realism with respect to motion is to be *completely eliminated.*

Now, we have seen that this naive realism is on occasions an essential part of our observational vocabulary. On these occasions (Paradigm I), the observation language contains the idea of the efficacy of *all* motion. Or, to express it in the material mode of speech, our experience in these situations is an experience of objects which move absolutely. Taking this into consideration, it is apparent that Galileo's proposal amounts to a partial revision of our observation language or of our experience. An

---

12. The idea that there is an absolute direction in the universe has a very interesting history. It rests on the structure of the gravitational field on the surface of the earth, or of that part of the earth which the observer knows, and generalizes the experiences made there. The generalization is only rarely regarded as a separate hypothesis; it rather enters the 'grammar' of common sense and gives the terms 'up' and 'down' an absolute meaning. (This is a 'natural interpretation', in precisely the sense that was explained in the text above.) Lactantius, a Church father of the fourth century, appeals to this meaning when he asks (*Divinae Institutiones*, 111, De Falsa Sapientia): 'Is one really going to be so confused as to assume the existence of humans whose feet are above their heads? Where trees and fruit grow not upwards, but downwards?' The same use of language is presupposed by that 'mass of untutored men' who raise the question why the antipodeans are not falling off the earth (Pliny, *Natural History*, II, pp. 161–6; see also Ptolemy, *Syntaxis*, I, 7). The attempts of Thales, Anaximenes and Xenophanes to find support for the earth which prevents it from falling 'down' (Aristotle, *De Coelo*, 294a12ff) shows that almost all early philosophers, with the sole exception of Anaximander, shared in this way of thinking. (For the Atomists, who assume that the atoms originally fall 'down', see Jammer, *Concepts of Space*, Cambridge, Mass., 1953, p. 11.) Even Galileo, who thoroughly ridicules the idea of the falling antipodes (*Dialogue*, op. cit., p. 331), occasionally speaks of the 'upper half of the moon', meaning that part of the moon 'which is invisible to us'. And let us not forget that some linguistic philosophers of today 'who are too stupid to recognize their own limitations' (Galileo, op. cit., p. 327) want to revive the absolute meaning of 'up–down' at least *locally.* Thus the power over the minds of his contemporaries of a primitive conceptual frame, assuming an anisotropic world, which Galileo had also to fight, must not be underestimated. For an examination of some aspects of British common sense at the time of Galileo, including astronomical common sense, see E.M.W. Tillyard, *The Elizabethan World Picture*, London, 1963. The agreement between popular opinion and the centrally symmetric universe is frequently asserted by Aristotle, e.g. in *De Coelo*, p. 308a23f.

experience which partly *contradicts* the idea of the motion of the earth is turned into an experience that *confirms* it, at least as far as 'terrestrial things' are concerned.[13] This is what *actually happens*. But Galileo wants to persuade us that no change has taken place, that the second conceptual system is already universally *known*, even though it is not universally *used*. Salviati, his representative in the Dialogue, his opponent Simplicio, and Sagredo the intelligent layman all connect Galileo's method of argumentation with Plato's theory of *anamnesis* – a clever tactical move, typically Galilean one is inclined to say. Yet we must not allow ourselves to be deceived about the revolutionary development that is actually taking place.

Paradigm I: Motion of compact objects in stable surroundings of great spatial extension – deer observed by the hunter.		*Paradigm II:* Motion of objects in boats, coaches and other moving systems.	
Natural interpretation: All motion is operative.		*Natural interpretation:* Only relative motion is operative.	
Falling stone	Motion of earth	Falling stone	Motion of Earth
proves	*predicts*	*proves*	*predicts*
↓	↓	↓	↓
Earth at rest	Oblique motion of stone	No *relative* motion between starting point and stone	No relative motion between starting point and stone

The resistance against the assumption that shared motion is non-operative was equated with the resistance which forgotten ideas exhibit towards the attempt to make them known. Let us accept this *interpretation* of the resistance! But let us not forget its *existence*. We must then admit that it restricts the use of the relativistic ideas, confining them to *part* of our everyday experience. *Outside* this part, i.e. in interplanetary space, they are 'forgotten' and therefore not active. But outside this part there is not complete chaos. Other concepts are used, among them those very same absolutistic concepts which derive from the first paradigm. We not only use them, we must also admit that they are entirely adequate.

---

13. *Dialogue*, op. cit., pp. 132 and 416.

No difficulties arise as long as one remains within the limits of the first paradigm. 'Experience', i.e. the totality of all facts from all domains, cannot force us to carry out the change which Galileo wants to introduce. The motive for a change must come from a different source.

It comes, first, from the desire to see 'the whole [correspond] to its parts with wonderful simplicity',[14] as Copernicus had already expressed himself. It comes from the 'typically metaphysical urge' for unity of understanding and conceptual presentation. And the motive for a change is connected, secondly, with the intention to make room from the motion of the earth, which Galileo accepts and is not prepared to give up. The idea of the motion of the earth is closer to the first paradigm than to the second, or at least it was at the time of Galileo. This gave strength to the Aristotelian arguments, and made them plausible. To eliminate the plausibility, it was necessary to subsume the first paradigm under the second, and to extend the relative notions to all phenomena. The idea of *anamnesis* functions here as a psychological crutch, as a lever which smoothes the process of subsumption by concealing its existence. As a result we are now ready to apply the relative notions not only to boats, coaches, birds, but to the 'solid and well-established earth' as a whole. And we have the impression that this readiness was in us all the time, although it took some effort to make it conscious. This impression is most certainly erroneous: it is the result of Galileo's propagandistic machinations. We would do better to describe the situation in a different way, as a change of our conceptual system. Or, because we are dealing with concepts which belong to natural interpretations, and which are therefore connected with sensations in a very direct way, we should describe it as a *change of experience* that allows us to accommodate the Copernican doctrine. It is this change which underlies the transition from the Aristotelian point of view to the epistemology of modern science.

---

14. Ibid., p. 341. Galileo quotes here from Copernicus' address to Pope Paul III in *De Revolutionibus*; see also Chapter 10 and the *Narratio Prima* (quoted from E. Rosen, *Three Copernican Treatises*, New York, 1959, p. 165): 'For all these phenomena appear to be linked most nobly together, as by a golden chain; and each of the planets, by its position, and order, and every inequality of its motion, bears witness that the earth moves and that we who dwell upon the globe of the earth, instead of accepting its changes of position, believe that the planets wander in all sorts of motions of their own.' Note that empirical reasons are absent from the argument and have to be, for Copernicus himself admits (*Commentariolus*, op. cit., p. 57) that the Ptolemaic theory is 'consistent with the numerical data'.

For experience now ceases to be the unchangeable fundament which it is both in common sense and in the Aristotelian philosophy. The attempt to support Copernicus makes experience 'fluid' in the very same manner in which it makes the heavens fluid, 'so that each star roves around in it by itself'.[15] An empiricist who starts from experience, and builds on it without ever looking back, now loses the very ground on which he stands. Neither the earth, 'the solid, well-established earth', nor the facts on which he usually relies can be trusted any longer. It is clear that a philosophy that uses such a fluid and changing experience needs new methodological principles which do not insist on an asymmetric judgement of theories by experience. *Classical physics* intuitively adopts such principles; at least its great and independent thinkers, such as Newton, Faraday, Boltzmann proceed in this way. But its *official doctrine* still clings to the idea of a stable and unchanging basis. The clash between this doctrine and the actual procedure is concealed by a tendentious presentation of the *results* of research that hides their revolutionary origin and suggests that they arose from a stable and unchanging source. These methods of concealment start with Galileo's attempt to introduce new ideas under the cover of anamnesis, and they culminate in Newton.[16] They must be exposed if we want to arrive at a better account of the progressive elements in science.

My discussion of the anti-Copernican argument is not yet complete. So far, I have tried to discover what assumption will make a stone *that moves alongside a moving tower* appear to fall 'straight down', instead of being seen to move in an arc. The assumption, which I shall call the *relativity principle*, that our senses notice only relative motion and are insensitive to a motion which objects have in common, was seen to do the trick. What remains to be explained is *why the stone stays with the tower* and is not left behind. In order to save the Copernican view, one must explain not only why a motion that preserves the relation among visible objects *remains unnoticed*, but also, why a common motion of various objects does not affect their relation. That is, one must explain why such a motion is not a causal agent. Turning the question around in the manner explained in the text to footnote 10, page 58 of the last chapter, it is now apparent that the anti-Copernican argument described there rests on *two* natural interpretations: viz, the *epistemological*

---

15. *Dialogue*, op. cit., p. 120.
16. 'Classical Empiricism', op. cit.

assumption that absolute motion is always *noticed*, and the *dynamical principle* that objects (such as the falling stone) which are not interfered with assume their natural motion. For Aristotelians the natural motion of an object not interfered with is *rest*, i.e. constancy of qualities and of position.[17] This corresponds to our own experience, where things have to be pushed around to move. The discovery of seeds, bacteria, viruses would have been impossible without a firm belief in the qualitative part of the law – and it confirmed it in a most impressive way. Using this law scientists inferred that a stone dropped from a tower situated on a moving earth would be left behind. Thus the relativity principle must be combined with a new law of inertia in such a fashion that the motion of the earth can still be asserted. One sees at once that the following law, the *principle of circular inertia* as I shall call it, provides the required solution: an object that moves with a given angular velocity on a frictionless sphere around the centre of the earth will continue moving with the same angular velocity for ever. Combining the appearance of the falling stone with the relativity principle, the principle of circular inertia and some simple assumptions concerning the composition of velocities,[18] we obtain an argument which no longer endangers Copernicus' view, but can be used to give it partial support.

The relativity principle was defended in two ways. The first was by showing how it helps Copernicus: this defence is *ad hoc* but not

---

17. This is the *general* account of motion. In the *cosmological* account we have circular motion above and up-and-down motions on earth.

18. These assumptions were not at all a matter of course, but conflicted with some very basic ideas of Aristotelian physics. The principle of circular inertia is related to the impetus theory, but not identical with it. The impetus theory retains the idea that it needs a force to bring about change, but it puts the force inside the changing object. Once pushed, an object continues moving in the same way in which a heated object stays warm – both contain the cause of their new state. Galileo modifies this idea in two ways. First, the circular motion is supposed to go on forever while an object kept moving by impetus will gradually slow down, just as a heated object, its analogue, gradually becomes colder. The argument for this modification is given in the text below; it is purely rhetorical. Secondly, the eternal circular motions must proceed without a cause: if relative motions are not operative, then introducing a motion with the same centre and the same angular velocity as a circular motion upheld by impetus cannot eliminate forces: we are on the way from impetus to momentum (cf. A. Maier, *Die Vorläufer Galileis im 14. Jahrhundert*, Rome, 1949). All these changes are overlooked by those who assume that the transition was the simple result of a new and better dynamics and that the dynamics was already available, but had not yet been applied in a determined way.

objectionable, because necessary for revealing natural interpretations. The second was by pointing to its function in common sense, and by surreptitiously generalizing that function (see above). No independent argument was given for its validity. Galileo's support for the principle of circular inertia is of exactly the same kind. He introduces the principle, again not by reference to experiment or to independent observation, but by reference to what everyone is already supposed to know.

> *Simplicio:* So you have not made a hundred tests, or even one? And yet you so freely declare it to be certain? . . .
>
> *Salviati:* Without experiment, I am sure that the effect will happen as I tell you, because it must happen that way; and I might add that you yourself also know that it cannot happen otherwise, no matter how you may pretend not to know it. . . . But I am so handy at picking people's brains that I shall make you confess this in spite of yourself.[19]

Step by step, Simplicio is forced to admit that a body that moves, without friction, on a sphere concentric with the centre of the earth will carry out a 'boundless', a 'perpetual' motion. We know, of course, especially after the analysis we have just completed of the non-operative character of shared motion, that what Simplicio accepts is based neither on experiment nor on corroborated theory. It is a daring new suggestion involving a tremendous leap of the imagination.[20] A little more analysis then shows that this suggestion is connected with experiments, such as the 'experiments' of the *Discorsi*[21] by *ad hoc* hypotheses. (The amount of

---

19. *Dialogue*, op. cit., p. 145.

20. For a Copernican the only leap involved was the identification of the earth as a celestial object. According to Aristotle celestial objects move in circles and 'a body that moves in a circle has neither heaviness nor lightness for it cannot change its distance from the centre, neither in a natural nor in a forced way'. *De Coelo*, 269b34f.

21. Incidentally, many of the 'experiences' or 'experiments' used in the arguments about the motion of the earth are entirely fictitious. Thus Galileo, in his *Trattato della Sfera* (*Opere*, Vol. II, pp. 211 ff), which 'follows the opinion of Aristotle and of Ptolemy' (p. 223), uses this argument against a rotation of the earth: '. . . objects which one lets fall from high places to the ground such as a stone from the top of a tower would not fall towards the foot of that tower; for during the time which the stone coming rectilinearly towards the ground, spends in the air, the earth, escaping it, and moving towards the east would receive it in a part far removed from the foot of the tower *in exactly the same manner in which a stone that is dropped from the mast of a rapidly moving ship will not fall towards its foot, but more towards the stern*' (p. 224). The italicized reference to the behaviour of stones on ships is again used in the *Dialogue* (p. 126),

friction to be eliminated follows not from independent investigations –
such investigations commence only much later, in the 18th century – but
from the result to be achieved, viz. the circular law of inertia.) Viewing
natural phenomena in this way leads to a re-evaluation of all experience,
as we have seen. We can now add that it leads to the invention of a *new
kind of experience* that is not only more sophisticated *but also far more
speculative than* the experience of Aristotle or of common sense. Speaking
paradoxically, but not incorrectly, one may say that *Galileo invents an
experience that has metaphysical ingredients.* It is by means of such an
experience that the transition from a geostatic cosmology to the point of
view of Copernicus and Kepler is achieved.[22]

---

when the Ptolemaic arguments are discussed, but it is no longer accepted as correct. 'It seems
to be an appropriate time', says Salviati (ibid., p. 180), 'to take notice of a certain generosity on
the part of the Copernicans towards their adversaries when, with perhaps too much liberality,
they concede as true and correct a number of experiments which their opponents have never
made. Such for example is that of the body falling from the mast of a ship while it is in motion.
. . .' Earlier, on p. 154, it is implied rather than observed that the stone will fall to the foot of
the mast, even if the ship should be in motion, while a possible experiment is discussed on
p. 186. Bruno (*La Cena de le Ceneri, Opere Italiane*, I, ed. Giovanni Gentile, Bari, 1907, p.
83) takes it for granted that the stone will arrive at the foot of the mast. It should be noted
that the problem did not readily lend itself to an experimental solution. Experiments were
made, but their results were far from conclusive. See A. Armitage, 'The Deviation of Falling
Bodies', *Annals of Science* 5, 1941–7, pp. 342ff, and A. Koyré, *Metaphysics and Measurement*,
Cambridge, 1968, pp. 89ff. The tower argument can be found in Aristotle, *De Coelo*, 296b22,
and Ptolemy, *Syntaxis*, i, 8. Copernicus discusses it in the same chapter of *De Revol*, but tries to
defuse it in the next chapter. Its role in the Middle Ages is described in M. Clagett, *The Science
of Mechanics in the Middle Ages*, Madison, 1959, Chapter 10.

22. Alan Chalmers, in an interesting and well-argued paper ('The Galileo that
Feyerabend Missed: An Improved Case Against Method' in J.A. Schuster and R.R. Yeo
(eds), *The Politics and Rhetoric of Scientific Method*, Dordrecht, 1986, pp. 1ff), distinguishes
'between Galileo's contributions to a new science, on the one hand, and the question of the
social conditions in which that science is developed and practised, on the other', admits
that 'propaganda' (though much less than I suggest) may have been part of his attempt to
change the latter, but emphasizes that it does not affect the former. 'The main source for
Galileo's contribution to science itself', says Chalmers, 'is his *Two New Sciences*'. This is
the work I should have studied to explore Galileo's procedure. But the *Two New Sciences*
do not deal with the topic I was discussing, viz. the transition to Copernicus. *Here* Galileo
used procedures rather different from those of his later work. Lynn Thorndike, who shares
Chalmers' evaluation of the *Dialogue*, wished that Galileo had written a systematic textbook
on that subject (*A History of Magic and Experimental Science*, Vol. 6, New York, 1941, pp. 7
and 62: 'Galileo might have done better to write a systematic textbook than his provocative
dialogues'). Now for such a textbook to have substance it would have to be as general as its

Aristotelian rival and it would have to show how and why Aristotelian concepts needed to be replaced at the most elementary level. Aristotelian concepts, though abstract, were closely related to common sense. Hence it was necessary to replace some common notions by others (I am now speaking about what Chalmers calls 'perceptual relativity' – p. 7). Two questions arise: How big were the changes? And was propaganda (rhetoric, were 'irrational moves') needed to carry them out? My answer to the latter question is that discourse attempting to bring about major conceptual changes is a normal part of science, common sense, and cultural exchange (for the latter cf. Chapter 16 and Chapter 17, item vi, 'open exchange'), and that it differs from the discourse carried out *within* a more or less stable framework. Personally, I am quite prepared to make it part of rationality. But there exist philosophical schools that oppose it or call it incoherent (see Chapter 10 of *Farewell to Reason* which discusses some of Hilary Putnam's views). *Using the terminology of these schools* I speak of Galileo's 'trickery', etc. And I add that science contains ingredients that occasionally need such 'trickery' to become acceptable. The difference between the *Sciences* and the *Dialogue*, therefore, is not between science and sociology but between technical changes in a narrow field and basic changes, realistically interpreted. My answer to the first question is that perceptual relativity, though acknowledged by many scholars (and by Aristotle himself), was not a common possession (Galileo points out that even some of his fellow scientists stumbled at this point) and thus had to be argued for. This is not at all surprising, as my discussion of qualitative difficulties in Chapter 5 shows. Besides, is it really true that a traveller on a boat sees the harbour as receding as if it were removed by some strange force? I conclude that Galileo's 'trickery' was necessary for a proper understanding of the new cosmology, that it is 'trickery' only for philosophies that set narrow conditions on conceptual change, and that it should be extended to areas still restricted by such conditions (in Chapter 12 I argue that the mind–body problem is one such area).

8

In addition to natural interpretations, Galileo also changes sensations that seem to endanger Copernicus. He admits that there are such sensations, he praises Copernicus for having disregarded them, he claims to have removed them with the help of the telescope. However, he offers no theoretical reasons why the telescope should be expected to give a true picture of the sky.

I repeat and summarize. An argument is proposed that refutes Copernicus by observation. The argument is inverted in order to discover the natural interpretations which are responsible for the contradiction. The offensive interpretations are replaced by others. Propaganda and appeal to distant, and highly theoretical, parts of common sense are used to defuse old habits and to enthrone new ones. The new natural interpretations, which are also formulated explicitly, as auxiliary hypotheses, are established partly by the support they give to Copernicus and partly by plausibility considerations and *ad hoc* hypotheses. An entirely new 'experience' arises in this way. There is as yet no independent evidence, but this is no drawback; it takes time to assemble facts that favour a new cosmology. For what is needed is a new dynamics that explains both celestial and terrestrial motions, a theory of solid objects, aero-dynamics; and all these sciences are still hidden in the future.[1] *But their task is now well-defined*, for Galileo's assumptions, his *ad hoc* hypotheses included, are sufficiently clear and simple to prescribe the direction of future research.

Let it be noted, incidentally, that Galileo's procedure drastically reduces the content of dynamics. Aristotelian dynamics was a general theory of change, comprising locomotion, qualitative change, generation and corruption, and it could also be applied to mental processes. Galileo's dynamics and its successors deal with *locomotion* only, and here again just with the locomotion of *matter*. Other kinds of motion are pushed

---

1. Galileo's circular law is not the right dynamics. It fits neither the epicycles which still occur in Copernicus, nor Kepler's ellipses. In fact, it is refuted by both. Still, Galileo regards it as an essential ingredient of the Copernican point of view and tries to remove bodies, such as comets, whose motion quite obviously is not circular, from interplanetary space. In his *Assayer* 'Galileo talked about comets [and interpreted them as illusions, similar to rainbows] in order to protect the Copernican system from possible falsifications.' P. Redondi, *Galileo: Heretic*, Princeton, 1987, pp. 145, 31.

aside with the promissory note (due to Democritos) that locomotion will eventually be capable of explaining all motion. Thus a comprehensive empirical theory is replaced by a narrow theory plus a metaphysics of motion,[2] just as an 'empirical' experience is replaced by an experience that contains speculative elements. *Counterinduction*, however, is now seen to play an important role both *vis-à-vis* theories and *vis-à-vis* facts. It clearly aids the advancement of science. This concludes the considerations begun in Chapter 6. I now turn to another part of Galileo's propaganda campaign, dealing not with natural interpretations but with the *sensory core* of our observational statements.

Replying to an interlocutor who expressed his astonishment at the small number of Copernicans, Salviati, who 'act[s] the part of

---

2. The so-called scientific revolution led to astounding discoveries and considerably extended our knowledge of physics, physiology, and astronomy. This was achieved by pushing aside and regarding as irrelevant, *and often as non-existent*, those facts which had supported the older philosophy. Thus the evidence for witchcraft, demonic possession, the existence of the devil, etc., was disregarded *together* with the 'superstitions' it once confirmed. The result was that 'towards the close of the Middle Ages science was forced away from human psychology, so that even the great endeavour of Erasmus and his friend Vives, as the best representatives of humanism, did not suffice to bring about a reapproachment, and psychopathology had to trail centuries behind the developmental trend of general medicine and surgery. As a matter of fact ... the divorcement of medical science from psychopathology was so definite that the latter was always totally relegated to the domain of theology and ecclesiastic and civil law – two fields which naturally became further and further removed from medicine. ...' G. Zilboorg, MD, *The Medical Man and the Witch*, Baltimore, 1935, pp. 3ff and 70ff. Astronomy advanced, but the knowledge of the human mind slipped back into an earlier and more primitive stage. Another example is astrology. 'In the early stages of the human mind,' writes A. Comte (*Cours de Philosophie Positive*, Vol. III, pp. 273–80, ed. Littré, Paris, 1836), 'these connecting links between astronomy and biology were studied from a very different point of view, *but at least* they were studied and not left out of sight, as is the common tendency in our own time, under the restricting influence of a nascent and incomplete positivism. Beneath the chimerical belief of the old philosophy in the physiological influence of the stars, there lay a strong, though confused recognition of the truth that the facts of life were in some way dependent on the solar system. Like all primitive inspirations of man's intelligence this feeling needed rectification by positive science, but not destruction; though unhappily in science, as in politics, it is often hard to reorganize without some brief period of overthrow.' A third area is mathematics. Aristotle had developed a highly sophisticated theory of the continuum that overcame the difficulties raised by Zeno and anticipated quantum theoretical ideas on motion (see footnote 15 and text of Chapter 5). Most physicists returned to the idea of a continuum consisting of indivisible elements – if they considered such recondite matters, that is.

Copernicus',[3] gives the following explanation: 'You wonder that there are so few followers of the Pythagorean opinion [that the earth moves] while I am astonished that there have been any up to this day who have embraced and followed it. Nor can I ever sufficiently admire the outstanding acumen of those who have taken hold of this opinion and accepted it as true: they have, through sheer force of intellect, done such violence to their own senses as to prefer what reason told them over that which sensible experience plainly showed them to be the contrary. For the arguments against the whirling [the rotation] of the earth we have already examined [the dynamical arguments discussed above] are very plausible, as we have seen; and the fact that the Ptolemaics and the Aristotelians and all their disciples took them to be conclusive is indeed a strong argument of their effectiveness. But the experiences which overtly contradict the annual movement [the movement of the earth around the sun] are indeed so much greater in their apparent force that, I repeat, there is no limit to my astonishment when I reflect that Aristarchus and Copernicus were able to make reason so conquer sense that, in defiance of the latter, the former became mistress of their belief.'[4]

A little later Galileo notes that 'they [the Copernicans] were confident of what their reason told them!'[5] And he concludes his brief account of the origins of Copernicanism by saying that 'with reason as his guide he [Copernicus] resolutely continued to affirm what sensible experience seemed to contradict'. 'I cannot get over my amazement', Galileo repeats, 'that he was constantly willing to persist in saying that Venus might go around the sun and might be more than six times as far from us at one time as at another, and still look always equal, when it should have appeared forty times larger.'[6]

The 'experiences which overtly contradict the annual movement', and which 'are much greater in their apparent force' than even the dynamical arguments above, consist in the fact that 'Mars, when it is

---

3. *Dialogue*, op. cit., pp. 131 and 256.

4. Ibid., p. 328. At other times Galileo speaks much more belligerently and dogmatically, and apparently without any awareness of the difficulties mentioned here. Cf. his preparatory notes for the letter to Grand Duchess Christina, *Opere*, V, pp. 367ff.

5. Ibid., p. 335.

6. Ibid., p. 339.

close to us ... would have to look sixty times as large as when it is most distant. Yet no such difference is to be seen. Rather, when it is in opposition to the sun and close to us it shows itself only four or five times as large as when, at conjunction, it becomes hidden behind the rays of the sun.' 'Another and greater difficulty is made for us by Venus which, if it circulates around the sun, as Copernicus says, would now be beyond it and now on this side of it, receding from and approaching towards us by as much as the diameter of the circle it describes. Then, when it is beneath the sun and very close to us, its disc ought to appear to us a little less than forty times as large as when it is beyond the sun and near conjunction. Yet the difference is almost imperceptible.'[7]

In an earlier essay, *The Assayer*, Galileo expressed himself still more bluntly. Replying to an adversary who had raised the issue of Copernicanism, he remarks that *'neither Tycho, nor other astronomers nor even Copernicus could clearly refute [Ptolemy]* inasmuch as a most important argument taken from the movement of Mars and Venus stood always in their way'. (This 'argument' is mentioned again in the *Dialogue*, and has just been quoted.) He concludes that 'the two systems' (the Copernican and the Ptolemaic) are 'surely false'.[8]

We see that Galileo's view of the origin of Copernicanism differs markedly from the more familiar historical accounts. He neither points to *new facts* which offer inductive *support* to the idea of the moving earth, nor does he mention any observations that would *refute* the geocentric point of view but be accounted for by Copernicanism. On the contrary, he emphasizes that not only Ptolemy, but Copernicus as well, is refuted by the facts,[9] and he praises Aristarchus and Copernicus for not having

---

7. Ibid., p. 334.

8. *The Assayer*, quoted from *The Controversy on the Comets of 1918*, op. cit., p. 185.

9. This refers to the period before the end of the 16th century; see Derek J. de S. Price, 'Contra-Copernicus: A Critical Re-Estimation of the Mathematical Planetary Theory of Ptolemy, Copernicus and Kepler', in M. Clagett (ed.), *Critical Problems in the History of Science*, Madison, 1959, pp. 197–218. Price deals only with the *kinematic* and the *optical* difficulties of the new views. (A consideration of the dynamical difficulties would further strengthen his case.) He points out that 'under the best conditions a geostatic or heliostatic system using eccentric circles (or their equivalents) with central epicycles can account for all angular motions of the planets to an accuracy better than 6′ ... excepting only the special theory needed to account for ... Mercury and excepting also the planet Mars which shows deviations up to 30′ from a theory. [This is] certainly better than the accuracy of 10′ which Copernicus himself stated as a satisfactory goal for his own theory', which was

given up in the face of such tremendous difficulties. He praises them for having proceeded *counterinductively*.

This, however, is not yet the whole story.

For while it might be conceded that Copernicus acted simply on faith, it may also be said that Galileo found himself in an entirely different position. Galileo, after all, invented a new dynamics. And he invented the telescope. The new dynamics, one might want to point out, removes the inconsistency between the motion of the earth and the 'conditions affecting ourselves and those in the air above us'.[10] And the telescope removes the 'even more glaring' clash between the changes in the apparent brightness of Mars and Venus as predicted on the basis of the Copernican scheme and as seen with the naked eye. This, incidentally, is also Galileo's own view. He admits that 'were it not for the existence of a superior and better sense than natural and common sense to join forces with reason' he would have been 'much more recalcitrant towards the Copernican system'.[11] The 'superior and better sense' is, of course, the *telescope*, and one is inclined to remark that the apparently counterinductive procedure was as a matter of fact induction (or conjecture plus refutation plus new conjecture), *but one based on a better experience*, containing not only better natural interpretations but also a better sensory core than was available to Galileo's Aristotelian predecessors.[12] This matter must now be examined in some detail.

---

difficult to test, especially in view of the fact that refraction (almost 1° on the horizon) was not taken into account at the time of Copernicus, and that the observational basis of the predictions was less than satisfactory.

Carl Schumacher (*Untersuchungen über die ptolemäische Theorie der unteren Planeten*, Münster, 1917) has found that the predictions concerning Mercury and Venus made by Ptolemy differ at most by an amount of 30′ from those of Copernicus. The deviations found between modern predictions and those of Ptolemy (and Copernicus), which in the case of Mercury may be as large as 7°, are due mainly to wrong constants and initial conditions, including an incorrect value of the constant of precession. For the versatility of the Ptolemaic scheme see N.R. Ilanson, *Isis*, No. 51, 1960, pp. 150–8.

10. Ptolemy, *Syntaxis*, i, 7.

11. *Dialogue*, op. cit., p. 328.

12. For this view see Ludovico Geymonat, *Galileo Galilei*, transl. Stillman Drake, New York, 1965 (first Italian edition 1957), p. 184. For the story of Galileo's invention and use of the telescope see R.S. Westfall, 'Science and Patronage', *Isis*, Vol. 76, 1985, pp. 11ff. According to Westfall, Galileo 'saw the telescope more as an instrument of patronage than as an instrument of astronomy' (p. 26) and had to be pushed into some astronomical applications by his pupil (and staunch Copernican) Castelli. Galileo's telescopes were better than others

The telescope is a 'superior and better sense' that gives new and more reliable evidence for judging astronomical matters. How is this hypothesis examined, and what arguments are presented in its favour?

In the *Sidereus Nuncius*,[13] the publication which contains his first telescopic observations, and which was also the first important contribution to his fame, Galileo writes that he 'succeeded [in building the telescope] through a deep study of the theory of refraction'. This suggests that he had *theoretical reasons* for preferring the results of telescopic observations to observations with the naked eye. But the particular reason he gives – his insight into the theory of refraction – is not *correct* and is not *sufficient* either.

The reason is not correct, for there exist serious doubts as to Galileo's knowledge of those parts of contemporary physical optics which were relevant for the understanding of telescopic phenomena. In a letter to Giuliano de Medici of 1 October 1610,[14] more than half a year after publication of the *Sidereus Nuncius*, he asks for a copy of Kepler's *Optics* of 1604,[15] pointing out that he had not yet been able to obtain it in Italy. Jean Tarde, who in 1614 asked Galileo about the construction of telescopes of pre-assigned magnification, reports in his diary that Galileo regarded the matter as a difficult one and that he had found

---

in circulation at the time and were much in demand. But he first satisfied the demands of potential patrons. Kepler, who complained about the quality of telescopes (see next chapter, footnote 21 and text) and who would have loved to possess a better instrument, had to wait.

13. *The Sidereal Messenger of Galileo Galilei*, transl. E. St Carlos, London, 1880, reissued by Dawsons of Pall Mall, 1960, p. 10.

14. Galileo, *Opere*, Vol. X, p. 441.

15. *Ad Vitellionem Paralipomena quibus Astronomiae Pars Optica Traditur*, Frankfurt, 1604, to be quoted from *Johannes Kepler, Gesammelte Werke*, Vol. II, Munich, 1939, ed. Franz Hammer. This particular work will be referred to as the 'optics of 1604'. It was the only useful optics that existed at the time. The reason for Galileo's curiosity was most likely the many references to this work in Kepler's reply to the *Sidereus Nuncius*. For the history of this reply as well as a translation see *Kepler's Conversation with Galileo's Sidereal Messenger*, transl. E. Rosen, New York, 1965. The many references to earlier work contained in the *Conversation* were interpreted by some of Galileo's enemies as a sign that 'his mask had been torn from his face' (G. Fugger to Kepler, 28 May 1610, Galileo, *Opere*, Vol. X, p. 361) and that he (Kepler) 'had well plucked him' – Maestlin to Kepler, 7 August (Galileo, *Opere*, Vol. X, p. 428). Galileo must have received Kepler's *Conversation* before 7 May (*Opere*, X, p. 349) and he acknowledges receipt of the printed *Conversation* in a letter to Kepler of 19 August (*Opere*, X, p. 421).

Kepler's *Optics* of 1611[16] so obscure 'that perhaps its own author had not understood it'.[17] In a letter to Liceti, written two years before his death, Galileo remarked that as far as he was concerned the nature of light was still in darkness.[18] Even if we consider such utterances with the care that is needed in the case of a whimsical author like Galileo, we must yet admit that his knowledge of optics was inferior by far to that of Kepler.[19] This is also the conclusion of Professor E. Hoppe, who sums up the situation as follows:

> Galileo's assertion that having heard of the Dutch telescope he reconstructed the apparatus by mathematical calculation must of course be understood with a grain of salt; for in his writings we do not find any calculations and the report, by letter, which he gives of his first effort says that no better lenses had been available; six days later we find him on the way to Venice with a better piece to hand it as a gift to the Doge Leonardi Donati. This does not look like calculation; it rather looks like trial and error. The calculation may well have been of a different kind, and here it succeeded, for on 25 August 1609 his salary was increased by a factor of three.[20]

---

16. *Dioptrice*, Augsburg, 1611, *Werke*, Vol. IV, Munich, 1941. This work was written after Galileo's discoveries. Kepler's reference to them in the preface has been translated by E. St Carlos, op. cit., pp. 37, 79ff. The problem referred to by Tarde is treated in Kepler's *Dioptrice*.

17. Geymonat, op. cit., p. 37.

18. Letter to Liceti of 23 June 1640, *Opere*, VIII, p. 208.

19. Kepler, the most knowledgeable and most lovable of Galileo's contemporaries, gives a clear account of the reasons why, despite his superior knowledge of optical matters, he 'refrained from attempting to construct the device'. 'You, however,' he addresses Galileo, 'deserve my praise. Putting aside all misgivings you turned directly to visual experimentation' (*Conversation*, op. cit., p. 18). It remains to add that Galileo, due to his lack of knowledge in optics, had no 'misgivings' to overcome: 'Galileo . . . was totally ignorant of the science of optics, and it is not too bold to assume that this was a most happy accident both for him and for humanity at large', Ronchi, *Scientific Change*, ed. Crombie, London, 1963, p. 550.

20. *Die Geschichte der Optik*, Leipzig, 1926, p. 32. Hoppe's judgement concerning the invention of the telescope is shared by Wolf, Zinner and others. Huyghens points out that superhuman intelligence would have been needed to invent the telescope on the basis of the available physics and geometry. After all, says he, we still do not understand the workings of the telescope. ('Dioptrica', *Hugenii Opuscula Postuma*, Ludg. Bat., 1903, 163, paraphrased after A.G. Kästner, *Geschichte der Mathematik*, Vol. IV, Göttingen, 1800, p. 60.)

Trial and error – this means that 'in the case of the telescope it was *experience* and not mathematics that led Galileo to a serene faith in the reliability of his device'.[21] This second hypothesis on the origin of the telescope is *also* supported by Galileo's testimony, in which he writes that he had tested the telescope 'a hundred thousand times on a hundred thousand stars and other objects'.[22] Such tests produced great and surprising successes. The contemporary literature – letters, books, gossip columns – testifies to the extraordinary impression which the telescope made as a means of improving *terrestrial vision*.

Julius Caesar Lagalla, Professor of Philosophy in Rome, describes a meeting of 16 April 1611, at which Galileo demonstrated his device: 'We were on top of the Janiculum, near the city gate named after the Holy Ghost, where once is said to have stood the villa of the poet Martial, now the property of the Most Reverend Malvasia. By means of this instrument, we saw the palace of the most illustrious Duke Altemps on the Tuscan Hills so distinctly that we readily counted its each and every window, even the smallest; and the distance is sixteen Italian miles. From the same place we read the letters on the gallery, which Sixtus erected in the Lateran for the benedictions, so clearly, that we distinguished even the periods carved between the letters, at a distance of at least two miles.'[23]

---

21. Geymonat, op. cit., p. 39.

22. Letter to Carioso, 24 May 1616, *Opere*, X, p. 357: letter to P. Dini, 12 May 1611, *Opere*, IX, p. 106: 'Nor can it be doubted that I, over a period of two years now, have tested my instrument (or rather dozens of my instruments) on hundreds and thousands of objects near and far, large and small, bright and dark; hence I do not see how it can enter the mind of anyone that I have simple-mindedly remained deceived in my observations.' The hundreds and thousands of experiments remind one of Hooke, and are most likely equally spurious. Cf. footnote 9 of Chapter 9.

23. Legalla, *De phaenomenis in orbe lunae novi telescopii usa a D. Galileo Galilei nunc iterum suscitatis physica disputatio* (Venice, 1612), p. 8; quoted from E. Rosen, *The Naming of the Telescope*, New York, 1947, pl. 54. The regular reports (*Avvisi*) of the Duchy of Urbino on events and gossip in Rome contain the following notice of the event: 'Galileo Galilei the mathematician, arrived here from Florence before Easter. Formerly a Professor at Padua, he is at present retained by the Grand Duke of Tuscany at a salary of 1,000 scudi. He has observed the motion of the stars with the *occiali*, which he invented or rather improved. Against the opinion of all ancient philosophers, he declares that there are four more stars or planets, which are satellites of Jupiter and which he calls the Medicean bodies, as well as two companions of Saturn. He has here discussed this opinion of his with Father Clavius, the Jesuit. Thursday evening, at Monsignor Malavasia's estate outside

Other reports confirm this and similar events. Galileo himself points to the 'number and importance of the benefits which the instrument may be expected to confer, when used by land or sea'.[24] The *terrestrial success* of the telescope was, therefore, assured. Its application to the *stars*, however, was an entirely different matter.

---

the St Pancratius gate, a high and open place, a banquet was given for him by Frederick Cesi, the marquis of Monticelli and nephew of Cardinal Cesi, who was accompanied by his kinsman, Paul Monaldesco. In the gathering there were Galileo; a Fleming named Terrentius; Persio, of Cardinal Cesi's retinue; [La] Galla, Professor at the University here; the Greek, who is Cardinal Gonzaga's mathematician; Piffari, Professor at Siena, and as many as eight others. Some of them went out expressly to perform this observation, and even though they stayed until one o'clock in the morning, they still did not reach an agreement in their views' (quoted from Rosen, op. cit., p. 31).

24. *Sidereal Messenger*, op. cit., p. ii. According to Berellus (*De Vero Telescopii Inventore*, Hague, 1655, p. 4), Prince Moritz immediately realized the military value of the telescope and ordered that its invention – which Berellus attributes to Zacharias Jansen – be kept a secret. Thus the telescope seems to have commenced as a secret weapon and was turned to astronomical use only later. There are many anticipations of the telescope to be found in the literature, but they mostly belong to the domain of natural magic and are used accordingly. An example is Agrippa von Nettesheim, who, in his book on occult philosophy (written 1509, Book II, Chapter 23), writes 'et ego novi ex illis miranda conficere, et specula in quibus quis videre poterit quaecunque voluerit a longissima distantia'. 'So may the toy of one age come to be the precious treasure of another', Henry Morley, *The Life of Cornelius Agrippa von Nettesheim*, Vol. II, p. 166.

9

Nor does the initial experience *with the telescope provide such reasons. The first telescopic observations of the sky are indistinct, indeterminate, contradictory and in conflict with what everyone can see with his unaided eyes. And the only theory that could have helped to separate telescopic illusions from veridical phenomena was refuted by simple tests.*

To start with, there is the problem of telescopic vision. This problem *is* different for celestial and terrestrial objects; and it was also *thought to be* different in the two cases.[1]

It was thought to be different because of the contemporary idea that celestial objects and terrestrial objects are formed from different materials and obey different laws. This idea entails that the result of an interaction of light (which connects both domains and has special properties) with terrestrial objects cannot, without further discussion, be extended to the sky. To this physical idea one added, entirely in accordance with the Aristotelian theory of knowledge (and also with present views about the matter), the idea that the senses are *acquainted* with the close appearance of terrestrial objects and are, therefore, able to perceive them distinctly, even if the telescopic image should be vastly distorted, or disfigured by coloured fringes. The stars are not known from close by.[2] Hence we cannot in their case use our *memory* for separating the contributions of the telescope and those which come from the object itself.[3] Moreover,

1. This is hardly ever realized by those who argue (with Kästner, op. cit., p. 133) that 'one does not see how a telescope can be good and useful on the earth and yet deceive in the sky'. Kästner's comment is directed against Horky. See below, text to footnotes 9–16 of the present chapter.

2. That the senses are acquainted with our everyday surroundings, but are liable to give misleading reports about objects outside this domain, is proved at once by the *appearance of the moon.* On the earth large but distant objects in familiar surroundings, such as mountains, are seen as being large, and far away. The appearance of the moon, however, gives us an entirely false idea of its distance and its size.

3. It is not too difficult to separate the letters of a familiar alphabet from a background of unfamiliar lines, even if they should happen to have been written with an almost illegible hand. No such separation is possible with letters which belong to an unfamiliar alphabet. The parts of such letters do not hang together to form distinct patterns which stand out from the background of general (optical) noise (in the manner described by K. Koffka, *Psychol. Bull.* 19, 1922, pp. 551ff, partly reprinted in M.D. Vernon (ed.), *Experiments in*

all the familiar cues (such as background, overlap, knowledge of nearby size, etc.), which constitute and aid our vision on the surface of the earth, are absent when we are dealing with the sky, so that new and surprising phenomena are bound to occur.[4] Only a new theory of vision, containing both hypotheses concerning the behaviour of light within the telescope and hypotheses concerning the reaction of the eye under exceptional circumstances, could have bridged the gulf between the heavens and the earth that was, and still is, such an obvious fact of physics and of astronomical observation.[5] We shall soon have occasion to comment on the theories that were available at the time and we shall see that they were unfit for the task and were refuted by plain and obvious facts. For the moment, I want to stay with the observations themselves and I want to comment on the contradictions and difficulties which arise when one tries to take the celestial results of the telescope at their face value, as indicating stable, objective properties of the things seen.

Some of these difficulties already announce themselves in a report of the contemporary *Avvisi*[6] which ends with the remark that 'even though they [the participants in the gathering described] went out expressly to perform this observation [of "four more stars or planets, which are satellites of Jupiter . . . as well as of two companions of Saturn"[7]], and even though they stayed until one in the morning, they still did not reach an agreement in their views.'

Another meeting that became notorious all over Europe makes the situation even clearer. About a year earlier, on 24 and 25 April 1610, Galileo had taken his telescope to the house of his opponent, Magini, in Bologna to demonstrate it to twenty-four professors of all faculties.

Visual Perception, London, 1966; see also the article by Gottschaldt in the same volume).

4. For the importance of cues such as diaphragms, crossed wires, background, etc., in the localization and shape of the telescope image and the strange situations arising when no cues are present see Chapter IV of Ronchi, *Optics*, op. cit., especially pp. 151, 174, 189, 191, etc. See also R.L. Gregory, *Eye and Brain*, New York, 1966, *passim* and p. 99 (on the autokinetic phenomenon). F.P. Kilpatrick (ed.), *Explorations in Transactional Psychology*, New York, 1961, contains ample material on what happens in the absence of familiar cues.

5. It is for this reason that the 'deep study of the theory of refraction' which Galileo pretended to have carried out (text to footnote 13 of Chapter 8) would have been quite *insufficient* for establishing the usefulness of the telescope; see also footnote 16 of the present chapter.

6. Details in Chapter 8, footnote 23.

7. This is how the ring of Saturn was seen at the time. See also R.L. Gregory, *The Intelligent Eye*, p. 119.

Horky, Kepler's overly-excited pupil, wrote on this occasion,[8] 'I never slept on the 24th or 25th April, day or night, but I tested the instrument of Galileo's in a thousand ways,[9] both on things here below and on those above. *Below it works wonderfully*; in the heavens it deceives one, as some fixed stars [Spica Virginis, for example, is mentioned, as well as a terrestrial flame] are seen double.[10] I have as witnesses most excellent men and noble doctors . . . and all have admitted the instrument to deceive. . . . This silenced Galileo and on the 26th he sadly left quite early in the morning . . . not even thanking Magini for his splendid meal. . . .' Magini wrote to Kepler on 26 May: 'He has achieved nothing, for more than twenty learned men were present; yet nobody has seen the new planets distinctly ['nemo perfecte vidit']; he will hardly be able to keep them.'[11] A few months later (in a letter signed by Ruffini) he repeats: 'Only some with sharp vision were convinced to some extent.'[12] After these and other negative reports had reached Kepler from all sides, like a paper avalanche, he asked Galileo for witnesses:[13] 'I do not want to hide it from you that quite a few Italians have sent letters to Prague asserting that they could not see those stars [the moons of Jupiter] with your own telescope. I ask myself how it can be that so many deny the phenomenon, including those who use a telescope. Now, if I consider what occasionally happens to me, then I do not at all regard it as impossible that a single person may see what thousands are unable to see. . . .'[14] Yet I regret that the confirmation

---

8. Galileo, *Opere*, Vol. X, p. 342 (my italics, referring to the difference commented upon above, between celestial and terrestrial observations).

9. The 'hundreds' and 'thousands' of observations, trials, etc., which we find here again are hardly more than a rhetorical flourish (corresponding to our 'I have told you a thousand times'). They cannot be used to infer a life of incessant observation.

10. Here again we have a case where external clues are missing. See Ronchi, *Optics*, op. cit., as regards the appearance of flames, small lights, etc.

11. Letter of 26 May, *Opere*, III.

12. Ibid., p. 196.

13. Letter of 9 August 1610, quoted from Caspar–Dyck, *Johannes Kepler in Seinen Briefen*, Vol. 1, Munich, 1930, p. 349.

14. Kepler, who suffered from Polyopia ('instead of a single small object at a great distance, two or three are seen by those who suffer from this defect. Hence, instead of a single moon ten or more present themselves to me', *Conversation*, op. cit., footnote 94; see also the remainder of the footnote for further quotations), and who was familiar with Platter's anatomical investigations (see S.L. Polyak, *The Retina*, Chicago, 1942, pp. 134ff for details and literature), was well aware of the need for a *physiological criticism of astronomical observations*.

by others should take so long in turning up. . . . Therefore, I beseech you, Galileo, give me witnesses as soon as possible. . . .' Galileo, in his reply of 19 August, refers to himself, to the Duke of Toscana, and Giuliano de Medici, 'as well as many others in Pisa, Florence, Bologna, Venice and Padua, who, however, remain silent and hesitate. Most of them are entirely unable to distinguish Jupiter, or Mars, or even the Moon as a planet. . . .'[15] – not a very reassuring state of affairs, to say the least.

Today we understand a little better why the direct appeal to telescopic vision was bound to lead to disappointment, especially in the initial stages. The main reason, one already foreseen by Aristotle, was that the senses applied under abnormal conditions are liable to give an abnormal response. Some of the older historians had an inkling of the situation, but they speak *negatively*, they try to explain the *absence* of satisfactory observational reports, the *poverty* of what is seen in the telescope.[16] They are unaware of the possibility that the observers might have been disturbed by *strong positive illusions* also. The extent of such illusions was not realized until quite recently, mainly as the result of the work of Ronchi and his school.[17] Here sizeable variations are reported in the

---

15. Caspar–Dyck, op. cit., p. 352.

16. Thus Emil Wohlwill, *Galileo und sein Kampf für die Kopernikanische Lehre*, Vol. 1, Hamburg, 1909, p. 288, writes: 'No doubt the unpleasant results were due to the lack of training in telescopic observation, and the restricted field of vision of the Galilean telescope as well as to the absence of any possibility for changing the distance of the glasses in order to make them fit the peculiarities of the eyes of the learned men. . . .' A similar judgement, though more dramatically expressed, is found in Arthur Koestler's *Sleepwalkers*, p. 369.

17. Cf. Ronchi, *Optics*, op. cit.: *Histoire de la Lumière*, Paris, 1956; *Storia del Cannochiale*, Vatican City, 1964; *Critica dei Fondamenti dell' Acustica e del'Ottica*, Rome, 1964; see also E. Cantore's summary in *Archives d'histoire des sciences*, December 1966, pp. 333ff. I would like to acknowledge at this place that Professor Ronchi's investigations have greatly influenced my thinking on scientific method. For a brief historical account of Galileo's work see Ronchi's article in A.C. Crombie (ed.), *Scientific Change*, London, 1963, pp. 542–61. How little this field is explored becomes clear from S. Tolansky's book *Optical Illusions*, London, 1964. Tolansky is a physicist who in his microscopic research (on crystals and metals) was distracted by one optical illusion after another. He writes: 'This turned our interest to the analysis of other situations, with the ultimate unexpected discovery that optical illusions can, and do, play a very real part in affecting many daily scientific observations. This warned me to be on the lookout and as a result I met more illusions than I had bargained for.' The 'illusions of direct vision', whose role in scientific research is slowly being rediscovered, were well known to mediaeval writers on optics,

placement of the telescopic image and, correspondingly, in the observed *magnification*. Some observers put the image right inside the telescope making it change its lateral position with the lateral position of the eye, exactly as would be the case with an after image, or a reflex inside the telescope – an excellent proof that one must be dealing with an 'illusion'.[18] Others place the image in a manner that leads to no magnification at all, although a linear magnification of over thirty may have been promised.[19] Even a doubling of images can be explained as the result of a lack of proper focusing.[20] Adding the many imperfections of the contemporary telescopes to these psychological difficulties,[21] one can well understand

---

who treated them in special chapters of their textbooks. Moreover, they treated lens-images as *psychological* phenomena, as results of a misapprehension, for an image 'is merely the appearance of an object outside its place' as we read in John Pecham (see David Lindberg, 'The "Perspectiva Communis" of John Pecham', *Archives Internationales d'histoire des sciences*, 1965, p. 51, as well as the last paragraph of Proposition ii/19 of Pecham's *Perspectiva Communis*, which is to be found in *John Pecham and the Science of Optics*, D. Lindberg (ed.), Madison, 1970, p. 171).

18. Ronchi, *Optics*, op. cit., p. 189. This may explain the frequently uttered desire to look *inside* the telescope. No such problems arise in the case of *terrestrial* objects whose images are regularly placed 'in the plane of the object' (ibid., p. 182).

19. For the magnification of Galileo's telescope see *The Sidereal Messenger*, op. cit., p. 11; see also A. Sonnefeld, 'Die Optischen Daten der Himmelsfernrohre von Galileo Galilei', *Jenaer Rundschau*, Vol. 7, 1962, pp. 207ff. The old rule 'that the size, position and arrangement according to which a thing is seen depends on the size of the angle through which it is seen' (R. Grosseteste, *De Iride*, quoted from Crombie, *Robert Grosseteste*, Oxford, 1953, p. 120), which goes back to Euclid, is *almost always wrong*. I still remember my disappointment when, having built a a reflector with an alleged linear magnification of about 150, I found that the moon was only about five times enlarged, and situated quite close to the ocular (1937).

20. The image remains sharp and unchanged over a considerable interval – the lack of focusing may show itself in a doubling, however.

21. The first usable telescope which Kepler received from Elector Ernst of Köln (who in turn had received it from Galileo), and on which he based his *Narratio de observatis a se quartuor Jovis satellibus*, Frankfurt, 1611, showed the stars as *squares* and intensely *coloured* (*Ges. Werke*, IV, p. 461). Ernst von Köln himself was unable to see anything with the telescope and he asked Clavius to send him a better instrument (*Archivio della Pontifica Universita Gregoriana*, 530, f 182r). Francesco Fontana, who from 1643 onwards observed the phases of Venus, notes an unevenness of the boundary (and infers mountains); see R. Wolf, *Geschichte der Astronomie*, Munich, 1877, p. 398. For the idiosyncrasies of contemporary telescopes and descriptive literature see Ernst Zinner, *Deutsche und Niederländische Astronomische Instrumente des 11 bis 18. Jahrhunderts*, Munich, 1956, pp. 216–21. Refer also to the author catalogue in the second part of the book.

the scarcity of satisfactory reports and one is rather astonished at the speed with which the reality of the new phenomena was accepted, and, as was the custom, publicly acknowledged.[22] This development becomes even more puzzling when we consider that many reports of even the best observers were either plainly *false*, and capable of being shown as such at the time, or else *self-contradictory*.

Thus Galileo reports unevenness, 'vast protuberances, deep chasms, and sinuosities'[23] at the inner boundary of the lighted part of the moon

---

22. Father Clavius (letter of 17 December 1610, *Opere*, X, p. 485), the astronomer of the powerful Jesuit Collegium Romanum, praises Galileo as the first to have observed the moons of Jupiter and he recognizes their reality. Magini, Grienberger, and others soon followed suit. It is clear that, in doing so, they did not proceed according to the methods prescribed by their own philosophy, or else they were very lax in the investigation of the matter. Professor McMullin (op. cit., footnote 32) makes much of this quick acceptance of Galileo's telescopic observations: 'The regular periods observed for the satellites and for the phases of Venus strongly indicated that they were not artefacts of physiology or optics. There was surely no need for "auxiliary sciences" . . .' 'There was no need for auxiliary sciences,' writes McMullin, while using himself the unexamined auxiliary hypothesis that astronomical events are distinguished from physiological events by their regularity and their intersubjectivity. But this hypothesis is *false*, as is shown by the moon illusion, the phenomenon of fata morgana, the rainbow, haloes, by the many microscopic illusions which are so vividly described by Tolansky, by the phenomena of witchcraft which survive in our textbooks of psychology and psychiatry, though under a different name, and by numerous other phenomena. The hypothesis was also *known to be false* by Pecham, Witelo, and other mediaeval scholars who had studied the regular and intersubjective 'illusions' created by lenses, mirrors, and other optical contrivances. In antiquity the falsehood of McMullin's hypothesis was *commonplace*. Galileo explicitly discusses and repudiates it in his book on comets. Thus a new theory of vision was needed, not just to *accept* the Galilean observations, but also to provide *arguments* for their astronomical reality. Of course, Clavius may not have been aware of this need. This is hardly surprising. After all, some of his sophisticated 20th-century successors, such as Professor McMullin, are not aware of it either. In addition we must point out that the 'regular periods' of the moons of Jupiter were not as well known as McMullin insinuates. For his whole life Galileo tried to determine these periods in order to find better ways of determining longitude at sea. He did not succeed. Later on the same problem returned in a different form when the attempt to determine the velocity of light with more than one moon led to conflicting results. This was found by Cassini shortly after Roemer's discovery – see I.B. Cohen, 'Roemer and the first determination of the velocity of light (1676)', *Isis*, Vol. 31 (1940), pp. 347ff. For the attitude of Clavius and the scientists of the Collegium Romanum see the very interesting book *Galileo in China* by Pasquale M. d'Elia, S.J., Cambridge, Mass., 1960. The early observations of the astronomers of the Collegium are contained in their own 'Nuncius Sidereus', *Opere*, III/I, pp. 291–8.

23. *The Sidereal Messenger*, op. cit., p. 8.

while the outer boundary 'appear[s] not uneven, rugged, and irregular, but perfectly round and circular, as sharply defined as if marked out with a pair of compasses, and without the indentations of any protuberances and cavities'.[24] The moon, then, seemed to be full of mountains at the inside but perfectly smooth at the periphery, and this despite the fact that the periphery *changed* as the result of the slight libration of the lunar body.[25] The moon and some of the planets, such as for example Jupiter,

---

24. Op. cit., p. 24. – see the drawing on p. 97 which is taken from Galileo's publication. Kepler in his *Optics* of 1604 writes (on the basis of observations with the unaided eye): 'It seemed as though something was missing in the circularity of the outmost periphery' (*Werke*, Vol. II, p. 219). He returns to this assertion in his *Conversation* (op. cit., pp. 28ff), criticizing Galileo's telescopic results by what he himself had seen with the unaided eye: 'You ask why the moon's outermost circle does not also appear irregular. I do not know how carefully you have thought about this subject or whether your query, as is more likely, is based on popular impression. For in my book [the *Optics* of 1604] I state that there was surely some imperfection in that outermost circle during full moon. Study the matter, and once again tell us, how it looks to you . . .' Here the results of naked eye observation are quoted against Galileo's telescopic reports – and with perfectly good reason, as we shall see below. The reader who remembers Kepler's polyopia (see footnote 14 to this chapter) may wonder how he could trust his senses to such an extent. The reply is contained in the following quotation (*Werke*, II, pp. 194ff): 'When eclipses of the moon begin, I, who suffer from this defect, become aware of the eclipse before all the other observers. Long before the eclipse starts, I even detect the direction from which the shadow is approaching, while the others, who have very acute vision, are still in doubt. . . . The afore-mentioned waviness of the moon [cf. the previous quotation] stops for me when the moon approaches the shadow, and the strongest part of the sun's rays is cut off . . .' Galileo has two explanations for the contradictory appearance of the moon. The one involves a lunar atmosphere (*Messenger*, op. cit., pp. 26ff). The other explanation (ibid., pp. 25ff), which involves the tangential appearance of series of mountains lying behind each other, is not really very plausible as the distribution of mountains near the visible side of the lunar globe does not show the arrangement that would be needed (this is now even better established by the publication of the Russian moon photograph of 7 October 1959; see Zdenek Kopal, *An Introduction to the Study of the Moon*, North Holland, 1966, p. 242).

25. The librations were noticed by Galileo. C.G. Righini, 'New Light on Galileo's Lunar Observations', in M.L. Righini-Bonelli and R. Shea (eds), *Reason, Experience and Mysticism in the Scientific Revolution*, New York, 1975, pp. 59ff. Thus it was not sloppiness of observations but the phenomena themselves that misguided Galileo.

In two letters to the journal *Science* (2 May and 10 October 1980) T.H. Whitaker accused me of giving a misleading account of Galileo's observational skill – I called him a poor observer when his lunar observations were in fact rather impressive. The accusation is refuted by the text to footnotes 29 and 30 and by footnote 46 of the present chapter.

were enlarged while the apparent diameter of the fixed stars decreased: the former were brought nearer whereas the latter were pushed away. 'The stars,' writes Galileo, 'fixed as well as erratic, when seen with the telescope, by no means appear to be increased in magnitude in the same proportion as other objects, and the Moon itself, gain increase of size; but in the case of the stars such increase appears much less, so that you may consider that a telescope, which (for the sake of illustration) is powerful enough to magnify other objects a hundred times, will scarcely render the stars magnified four or five times.'[26]

The strangest features of the early history of the telescope emerge, however, when we take a closer look at Galileo's *pictures of the moon*.

It needs only a brief look at Galileo's drawings, and a photograph of similar phases, to convince the reader that 'none of the features recorded

---

Whitaker obviously thought my quotations from Wolf (text to footnote 28) reflected my own opinion. He also points out that the copperplates of Galileo's observations are much better, from a modern point of view, than the woodcuts which accompanied the *Nuncius*. This is true but does not invalidate my description of the debate, which was based on the published account.

26. *Messenger*, op. cit., p. 38. See also the more detailed account in *Dialogue*, op. cit., pp. 336ff. 'The telescope, as it were, removes the heavens from us,' writes A. Chwalina in his edition of *Kleomedes, Die Kreisbewegung der Gestirne* (Leipzig, 1927, p. 90), commenting on the decrease of the apparent diameter of *all* stars with the sole exception of the sun and the moon. Later on, the different magnification of planets (or comets) and fixed stars was used as a means of distinguishing them. 'From experience, I know', writes Herschel in the paper reporting his first observation of Uranus (*Phil. Trans.*, 71, 1781, pp. 493ff – the planet is here identified as a *comet*), 'that the diameters of the fixed stars are not proportionally magnified with higher powers, as the planets are; therefore, I now put on the powers of 460 and 932, and found the diameter of the comet increased in proportion to the power, as it ought to be. . . .' It is noteworthy that the rule did not invariably apply to the telescopes in use at Galileo's time. Thus, commenting on a comet of November 1618, Horatio Grassi ('On the Three Comets of 1618', in *The Controversy of the Comets of 1618*, op. cit., p. 17) points out 'that when the comet was observed through a telescope it suffered scarcely any enlargement', and he infers, perfectly in accordance with Herschel's 'experience', that 'it will have to be said that it is more remote from us than the moon. . . .' In his *Astronomical Balance* (ibid., p. 80) he repeats that, according to the common experience of 'illustratious astronomers' from 'many parts of Europe' the comet observed with a very extended telescope received scarcely any increment. . . . Galileo (ibid., p. 177) accepts this as a fact, criticizing only the conclusions which Grassi wants to draw from it. All these phenomena refute Galileo's assertion (*Assayer*, op. cit., p. 204) that the telescope 'works always in the same way'. They also undermine his theory of irradiation (see footnote 56 to this chapter).

... can be safely identified with any known markings of the lunar landscape'.[27] Looking at such evidence it is very easy to think that 'Galileo was not a great astronomical observer; or else that the excitement of so many telescopic discoveries made by him at that time had temporarily blurred his skill or critical sense'.[28]

Now this assertion may well be true (though I rather doubt it in view of the quite extraordinary observational skill which Galileo exhibits on other occasions).[29] But it is poor in content and, I submit, not very interesting. No new suggestions emerge for additional research, and the possibility of a *test* is rather remote.[30] There are, however, other hypotheses which do lead to new suggestions and which show us how complex the situation was at the time of Galileo. Let us consider the following two.

Hypothesis I. Galileo recorded faithfully what he saw and in this way left us evidence of the shortcomings of the first telescopes as well as of the peculiarities of contemporary telescopic vision. Interpreted in this way Galileo's drawings are reports of exactly the same kind as are the reports emerging from the experiments of Stratton, Ehrismann, and Kohler[31] – except that the characteristics of the physical apparatus and

---

27. Kopal, op. cit., p. 207.

28. R. Wolf (*Geschichte der Astronomie*, p. 396) remarks on the poor quality of Galileo's drawings of the moon ('... seine Abbildung des Mondes kann man ... kaum ... eine Karte nennen'), while Zinner (*Geschichte der Sternkunde*, Berlin, 1931, p. 473) calls Galileo's observations of the moon and Venus 'typical for the observations of a beginner'. His picture of the moon, according to Zinner, 'has no similarity with the moon' (ibid., p. 472). Zinner also mentions the much better quality of the almost simultaneous observations made by the Jesuits (ibid., p. 473), and he finally asks whether Galileo's observations of the moon and Venus were not the result of a fertile brain, rather than of a careful eye ('sollte dabei ... der Wunsch der Vater der Beobachtung gewesen sein?') – a pertinent question, especially in view of the phenomena briefly described in footnote 34 to this chapter.

29. The discovery and identification of the moons of Jupiter were no mean achievements, especially as a useful stable support for the telescope had not yet been developed.

30. The reason, among other things, is the great variation of telescopic vision from one observer to the next; see Ronchi, *Optics*, op. cit., Chapter IV.

31. For a survey and some introductory literature see Gregory, op. cit., Chapter 11. For a more detailed discussion and literature, see K.W. Smith and W.M. Smith, *Perception and Motion*, Philadelphia, 1962, reprinted in part in M.D. Vernon, op. cit. The reader should also consult Ames' article 'Aniseikonic Glasses', *Explorations in Transactional Psychology*, which deals with the change of *normal* vision caused by only slightly abnormal optical conditions. A comprehensive account is given by I. Rock, *The Nature of Perceptual Adaptation*, New York, 1966.

the unfamiliarity of the objects seen must be taken into account too.[32] We must also remember the many conflicting views which were held about the surface of the moon, even at Galileo's time,[33] and which may have influenced what observers saw.[34] What would be needed in order to shed more light on the matter is an empirical collection of all the early telescopic results, preferably in parallel columns, including whatever pictorial representations have survived.[35] Subtracting instrumental peculiarities, such a collection adds fascinating material to a yet-to-be-written history of perception (and of science).[36] This is the content of Hypothesis I.

Hypothesis II is more specific than Hypothesis I, and develops it in a certain direction. I have been considering it, with varying degrees of enthusiasm, for the last two or three years and my interest in it has

---

32. Many of the old instruments, and excellent descriptions of them, are still available. See Zinner, *Deutsche und Niederlandische astronomische Instrumente*.

33. For interesting information the reader should consult the relevant passages of Kepler's *Conversation* as well as of his *Somnium* (the latter is now available in a new translation by E. Rosen, who has added a considerable amount of background material: *Kepler's Somnium*, ed. Rosen, Madison, 1967). The standard work for the beliefs of the time is still Plutarch's *Face on the Moon* (it will be quoted from H. Cherniss' translation of *Moralia XII*, London, 1967).

34. 'One describes the moon after objects one thinks one can perceive on its surface' (Kästner, op. cit., Vol. IV, p. 167, commenting on Fontana's observational reports of 1646). 'Maestlin even saw rain on the moon' (Kepler, *Conversation*, op. cit., pp. 29f, presenting Maestlin's own observational report); see also da Vinci, notebooks, quoted from J.P. Richter, *The Notebooks of Leonardo da Vinci*, Vol. II, New York, 1970, p. 167: 'If you keep the details of the spots of the moon under observation you will often find great variation in them, and this I myself have proved by drawing them. And this is caused by the clouds that rise from the waters in the moon. . . .' For the instability of the images of unknown objects and their dependence on belief (or 'knowledge') see Ronchi, *Optics*, op. cit., Chapter IV.

35. Chapter 15 of Kopal, op. cit., contains an interesting collection of exactly this kind. Wider scope has W. Schulz, *Die Anschauung vom Monde und seinen Gestalten in Mythos und Kunst der Völker*, Berlin, 1912.

36. One must, of course, also investigate the dependence of what is seen on the current methods of pictorial representation. Outside astronomy this was done by E. Gombrich, *Art and Illusion*, London, 1960, and L. Choulant, *A History and Bibliography of Anatomical Illustration*, New York, 1945 (translated, with additions, by Singer and others), who deals with anatomy. Astronomy has the advantage that *one* side of the puzzle, viz. the stars, is fairly simple in structure (much simpler than the uterus, for example) and relatively well known; see also Chapter 16, below.

been revived by a letter from Professor Stephen Toulmin, to whom I am grateful for his clear and simple presentation of the view. It seems to me, however, that the hypothesis is confronted by many difficulties and must, perhaps, be given up.

Hypothesis II, just like Hypothesis I, approaches telescopic reports from the point of view of the theory of perception; but it adds that the practice of telescopic observation and acquaintance with the new telescopic reports changed not only what was seen through the telescope, *but also what was seen with the naked eye*. It is obviously of importance for our evaluation of the contemporary attitude towards Galileo's reports.

That the appearance of the stars, and of the moon, may at some time have been much more indefinite than it is today was originally suggested to me by the existence of various theories about the moon which are incompatible with what everyone can plainly see with his own eyes. Anaximander's theory of partial stoppage (which aimed to explain the phases of the moon), Xenophanes' belief in the existence of different suns and different moons for different zones of the earth, Heraclitus' assumption that eclipses and phases are caused by the turning of the basins, which for him represented the sun and the moon[37] – all these views run counter to the existence of a stable and plainly visible surface, a 'face' such as we 'know' the moon to possess. The same is true of the theory of Berossos which occurs as late as Lucretius[38] and, even later, in Alhazen.

Now such disregard for phenomena which for us are quite obvious may be due either to a certain indifference towards the existing evidence, which was, however, as clear and as detailed as it is today, *or else to a difference in the evidence itself*. It is not easy to choose between these alternatives. Having been influenced by Wittgenstein, Hanson, and others, I was for some time inclined towards the second version, but it now seems to me that it is ruled out both by physiology (psychology)[39] and by historical

---

37. For these theories and further literature cf. J.L.D. Dreyer, *A History of Astronomy from Thales to Kepler*, New York, 1953.

38. For Berossos, see Toulmin's article in *Isis*, No. 38, 1967, p. 65. Lucretius writes (*On the Nature of Things*, transl. Leonard, New York, 1957, p. 216): 'Again, she may revolve upon herself/like to a ball's sphere – if perchance to be – / one half of her dyed o'er with glowing light / and by the revolution of that sphere / she may beget for us her varying shapes / until she turns that fiery part of her / full to the sight and open eyes of men. . . .'

39. See text to footnotes 50ff of my 'Reply to Criticism', op. cit., p. 246.

information. We need only remember how Copernicus disregarded the difficulties arising from the variations in the brightness of Mars and Venus, which were well known at the time.[40] And as regards the face of the moon, we see that Aristotle refers to it quite clearly when observing that 'the stars do not *roll*. For rolling involves rotation: but the "face", as it is called, of the moon is always seen.'[41] We may infer, then, that the occasional disregard for the stability of the face was due not to a lack of clear impressions, but to some widely held views about the unreliability of the senses. The inference is supported by Plutarch's discussion of the matter which plainly deals not with what is seen (except as evidence for or against certain views) but with certain *explanations* of phenomena otherwise *assumed to be well known*:[42] 'To begin with,' he says, 'it is absurd to call the figure seen in the moon an affection of vision . . . a condition which we call bedazzlement [glare]. Anyone who asserts this does not observe that this phenomenon should rather have occurred in relation to the sun, since the sun lights upon us keen and violent, and moreover does not explain why dull and weak eyes discern no distinction of shape in the moon but her orb for them has an even and full light whereas those of keen and robust vision make out more precisely and distinctly the pattern of facial features and more clearly perceive the variations.' 'The unevenness also entirely refutes the hypothesis,' Plutarch continues,[43] 'for the shadow that one sees is not continuous and confused, but is not badly depictured by the words of Agesianax: "She gleams with fire encircled, but within / Bluer than lapis

---

40. In antiquity the differences in the magnitudes of Venus and Mars were regarded as being 'obvious to our eyes', Simplicius, *De Coelo*, II, 12, Heiberg, p. 504. Polemarchus here considers the difficulties of Eudoxos' theory of homocentric spheres, viz. that Venus and Mars 'appear in the midst of the retrograde movement many times brighter, so that [Venus] on moonless nights causes bodies to throw shadows' (objection of Autolycus) and he may well be appealing to the possibility of a deception of the senses (which was frequently discussed by ancient schools). Aristotle, who must have been familiar with all these facts, does not mention them anywhere in *De Coelo* or in the *Metaphysics*, though he gives an account of Eudoxos' system and of the improvements of Polemarchus and Kalippus. See footnote 7 of Chapter 8.

41. *De Coelo*, 290a25ff.

42. Op. cit., p. 37. See also S. Sambursky, *The Physical World of the Greeks*, New York, 1962, pp. 244ff.

43. Ibid. However, footnote 17 to this chapter, Pliny's remark (*Hist. Nat.*, II, 43, 46) that the moon is 'now spotted and then suddenly shining clear', as well as da Vinci's report, referred to in footnote 34 to this chapter.

show a maiden's eye / And dainty brow, a visage manifest." In truth, the dark patches submerge beneath the bright ones which they encompass . . . and they are thoroughly entwined with each other so as to make the delineation of the figure resemble a painting.' Later on the stability of the face is used as an argument against theories which regard the moon as being made of fire, or air, for 'air is tenuous and without configuration, and so it naturally slips and does not stay in place'.[44] The *appearance* of the moon, then, seemed to be a well-known and distinct phenomenon. What was in question was the *relevance* of the phenomenon for astronomical theory.

We can safely assume that the same was true at the time of Galileo.[45]

But then we must admit that Galileo's observations could be checked with the naked eye and could in this way be exposed as illusory.

Thus the circular monster below the centre of the disk of the moon[46] is well above the threshold of naked-eye observation (its diameter is

---

44. Ibid., p. 50.

45. A strong argument *in favour* of this contention is Kepler's description of the moon in his *Optics* of 1604: he comments on the broken character of the boundary between light and shadow (*Werke*, II, p. 218) and describes the dark part of the moon during an eclipse as looking like torn flesh or broken wood (ibid., p. 219). He returns to these passages in the *Conversation* (op. cit., p. 27), where he tells Galileo that 'these very acute observations of yours do not lack the support of even my own testimony. For [in my] *Optics* you have the half moon divided by a wavy line. From this fact I deduced peaks and depressions in the body of the moon. [Later on] I describe the moon during an eclipse as looking like torn flesh or broken wood, with bright streaks penetrating into the region of the shadow.' Remember also that Kepler criticizes Galileo's telescopic reports on the basis of his own naked eye observations; see footnote 24 of this chapter.

46. 'There is one other point which I must on no account forget, which I have noticed and rather wondered at it. It is this: The middle of the Moon, as it seems, is occupied by a certain cavity larger than all the rest, and in shape perfectly round. I have looked at this depression near both the first and the third quarters, and I have represented it as well as I can in the second illustration already given. It produces the same appearance as to effects of light and shade as a tract like Bohemia would produce on the Earth, if it were shut in on all sides by very lofty mountains arranged on the circumference of a perfect circle; for the tract in the moon is walled in with peaks of such enormous height that the furthest side adjacent to the dark portion of the moon is seen bathed in sunlight before the boundary between light and shade reaches half way across the circular space . . .' (*Messenger*, op. cit., pp. 21ff). This description, I think, definitely refutes Kopal's conjecture of observational laxity. It is interesting to note the difference between the woodcuts in the *Nuncius* (p. 131, Figure 1) and Galileo's original drawing. The woodcut corresponds quite closely to the description while the original drawing with its impressionistic features ('Kaum eine Karte,' says Wolf) is vague enough to escape the accusation of gross observational error.

larger than 3½ minutes of arc), while a single glance convinces us that the face of the moon is not anywhere disfigured by a blemish of this kind. It would be interesting to see what contemporary observers had to say on the matter[47] or, if they were artists, what they had to draw on the matter.

I summarize what has emerged so far.

Galileo was only slightly acquainted with contemporary optical *theory*.[48] His telescope gave surprising results on the earth, and these results were duly praised. Trouble was to be expected in the sky, as we know now. Trouble promptly arose: the telescope produced spurious and contradictory phenomena and some of its results could be refuted by a simple look with the unaided eye. Only a new *theory* of telescopic vision could bring order into the chaos (which may have been still larger, due to the different phenomena seen at the time even with the naked eye) and could separate appearance from reality. Such a theory was developed by Kepler, first in 1604 and then again in 1611.[49]

According to Kepler, the place of the image of a punctiform object is found by first tracing the path of the rays emerging from the object according to the laws of (reflection and) refraction until they reach the eye, and by then using the principle (still taught today) that 'the image will be seen in the point determined by the backward intersection of the rays of vision from both eyes'[50]

---

47. 'I cannot help wondering about the meaning of that large circular cavity in what I usually call the left corner of the mouth,' writes Kepler (*Conversation*, op. cit., p. 28), and then proceeds to make conjectures as to its origin (conscious efforts by intelligent beings included).

48. Contemporary academic optics went beyond simple geometrical constructions (which Galileo may have known) and included an account of *what is seen* when looking at a mirror, or through a lens, or a combination of lenses. Excepting irradiation Galileo nowhere considers the properties of telescopic *vision*. Aristotelians writing after Galileo's telescopic observations did. See Redondi, op. cit., pp. 169ff.

49. I have here disregarded the work of della Porta (*De Refractione*) and of Maurolycus, who both anticipated Kepler in certain respects (and are duly mentioned by him). Maurolycus makes the important step [*Photismi de Lumine*, transl. Henry Crew, New York, 1940, p. 45 (on mirrors) and p. 74 (on lenses)] of considering only the cusp of the caustic; but a connection with what is seen on *direct* vision is still not established. For the difficulties which were removed by Kepler's simple and ingenious hypothesis see Ronchi, *Histoire de la Lumière*, Chapter III.

50. *Werke*, II, p. 72. The *Optics* of 1604 has been partly translated into German by F. Plehn, *J. Keplers Grundlagen der geometrischen Optik*, Leipzig, 1922. The relevant passages occur in section 2 of Chapter 3, pp. 38–48.

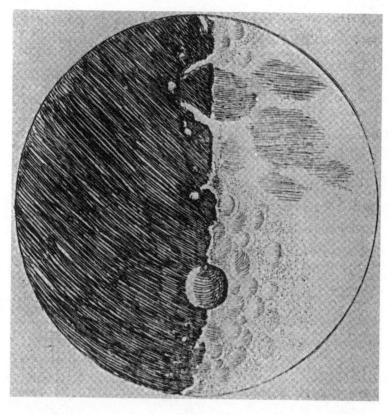

FIGURE 1. The shape of a lunar mountain and a walled plain, from Galileo *Sidereus Nuncius*, Venice, 1610 (cf. p. 109).

or, in the case of monocular vision, from the two sides of the pupil.[51] This rule, which proceeds from the assumption that 'the image is the work of the act of vision', is partly empirical and partly geometrical. It bases the position of the image on a 'metrical triangle'[52] or a 'telemetric triangle'[53]

51. Ibid., p. 67.

52. 'Cum imago sit visus opus', ibid., p. 64. 'In visione tenet sensus communis oculorum suorum distantiam ex assuefactione, angulos vero ad illam distantiam notat ex sensu contortionis oculorum', ibid., p. 66.

53. 'Triangulum distantiae mensorium', ibid., p. 67.

as Ronchi calls it,[54] that is constructed out of the rays which finally arrive at the eye and is used by the eye *and the mind* to place the image at the proper distance. Whatever the optical system, whatever the total path of the rays from the object to the observer, the rule says that the mind of the observer utilizes its *very last part only*, and bases its visual judgement, the perception, on it.

The rule considerably simplified the science of optics. However, it needs only a second to show that it is false: take a magnifying glass, determine its focus, and look at an object close to it. The telemetric triangle now reaches beyond the object to infinity. A slight change of distance brings the Keplerian image from infinity to close by and back to infinity. No such phenomenon is ever observed. We see the image, slightly enlarged, in a distance that is most of the time identical with the actual distance between the object and the lens. The visual distance of the image remains constant, however much we may vary the distance between lens and object and even when the image becomes distorted and, finally, diffuse.[55]

------

54. *Optics*, op. cit., p. 44. One should also consult the second chapter of this book for the history of pre-Keplerian optics.

55. Ibid., pp. 182, 202. This phenomenon was known to everyone who had used a magnifying glass only once, Kepler included. Which shows that disregard of familiar phenomena does not entail that the phenomena were seen differently (see text to footnote 44 to this chapter). Isaac Barrow's account of the difficulty of Kepler's rule was mentioned above (text to footnote 18 to Chapter 5). According to Berkeley (op. cit. p. 141) 'this phenomenon . . . entirely subverts the opinion of those who will have us judge of distances by lines and angles . . .' Berkeley replaces this opinion by his own theory according to which the mind judges distances from the clarity or confusion of the primary impressions. Kepler's idea of the telemetric triangle was adopted at once by almost all thinkers in the field. It was given a fundamental position by Descartes according to whom 'Distantiam . . . discimus, per mutuam quandam conspirationem oculorum' (*Dioptrice*, quoted from *Renati Descartes Specima Philosophiae*, Amsterdam, 1657, p. 87). 'But,' says Barrow, 'neither this nor any other difficulty shall . . . make me renounce that which I know to be manifestly agreeable to reason.' It is this attitude which was responsible for the slow advance of a scientific theory of *eye glasses* and of visual optics in general. 'The reason for this peculiar phenomenon,' writes Moritz von Rohr (*Das Brillenglas als optisches Instrument*, Berlin, 1934, p. 1), 'is to be sought in the close connection between the eye glass and the eye and it is impossible to give an acceptable theory of eye glasses without understanding what happens in the process of vision itself . . .' The telemetric triangle omits precisely this process, or rather gives a simplistic and false account of it. The state of optics at the beginning of the 20th century is well described in A. Gullstrand's 'Appendices to Part I' of Helmholtz's *Treatise on Physiological Optics*, transl. Southall, New York, 1962, pp. 261ff. We read here how a return to the phycho-physiological process

This, then, was the actual situation in 1610 when Galileo published his telescopic findings. How did Galileo react to it? The answer has already been given: he raised the telescope to the state of a 'superior and better sense'.[56]

---

of vision enabled physicists to arrive at a more reasonable account even of the physics of optical imagery: 'The reason why the laws of actual optical imagery have been, so to speak, summoned to life by the requirements of physiological optics is due partly to the fact that by means of trigonometrical calculations, tedious to be sure, but easy to perform, it has been possible for the optical engineer to get closer to the realities of his problem. Thus, thanks to the labours of such men as Abbe and his school, technical optics has attained its present splendid development; whereas, with the scientific means available, a comprehensive grasp of the intricate relations in the case of the imagery in the eye has been actually impossible.'

56. 'O Nicholas Copernicus, what a pleasure it would have been for you to see this part of your system confirmed by so clear an experiment!' writes Galileo, implying that the new telescopic phenomena are additional support for Copernicus (*Dialogue*, op. cit., p. 339). The difference in the appearance of planets and fixed stars (see footnote 26 to this chapter) he explains by the hypothesis that 'the very instrument of seeing [the eye] introduces a hindrance of its own' (ibid., p. 335), and that the telescope removes this hindrance, viz. *irradiation*, permitting the eye to see the stars and the planets as they really are. (Mario Giuducci, a follower of Galileo, ascribed irradiation to refraction by moisture on the surface of the eye, *Discourse on the Comets of 1618*, op. cit., p. 47.) This explanation, plausible as it may seem (especially in view of Galileo's attempt to show how irradiation can be removed by means other than the telescope) is not as straightforward as one might wish. Gullstrand (op. cit., p. 426) says that 'owing to the properties of the wave surface of the bundle of rays refracted in the eye . . . it is a mathematical impossibility for any cross section to cut the caustic surface in a smooth curve in the form of a circle concentric with the pupil'. Other authors point to 'inhomogeneities in the various humours, and above all in the crystalline lens' (Ronchi, *Optics*, op. cit., p. 104). Kepler gives this account (*Conversation*, op. cit., pp. 33ff): 'Point sources of light transmit their cones to the crystalline lens. There refraction takes place, and behind the lens the cones again contract to a point. But this point does not reach as far as the retina. Therefore, the light is dispersed once more, and spreads over a small area of retina, whereas it should impinge on a point. Hence the telescope, by introducing another refraction, makes this point coincide with the retina. . . .' Polyak, in his classical work *The Retina*, attributes irradiation partly to 'defects of the dioptrical media and to the imperfect accommodation' but 'chiefly' to the 'peculiar structural constitution of the retina itself' (p. 176), adding that it may be a function of the brain also (p. 429). None of these hypotheses covers *all* the facts known about irradiation. Gullstrand, Ronchi, and Polyak (if we omit his reference to the brain which can be made to explain anything we want) cannot explain the disappearance of irradiation in the telescope. Kepler, Gullstrand and Ronchi also fail to give an account of the fact, emphasized by Ronchi, that large objects show no irradiation at their edges ('Anyone undertaking to account for the phenomenon of irradiation must admit that when he looks at an electric bulb from afar so that it seems like a point, he sees it surrounded by an immense crown of rays whereas from nearby he sees

What were his reasons for doing so? This question brings me back to the problems raised by the evidence (against Copernicus) that was reported and discussed in Chapter 8.

---

nothing at all around it,' *Optics*, op. cit., p. 105). We know now that large objects are made definite by the lateral inhibitory interaction of retinal elements (which is further increased by brain function), see Ratliff, *Mach Bands*, p. 146, but the variation of the phenomenon with the diameter of the object and under the conditions of telescopic vision remains unexplored. Galileo's hypothesis received support mainly from its agreement with the Copernican point of view and was, therefore, largely *ad hoc*.

10

On the other hand, there are some telescopic phenomena which are plainly Copernican. Galileo introduces these phenomena as independent evidence for Copernicus while the situation is rather that one refuted view – Copernicanism – has a certain similarity to phenomena emerging from another refuted view – the idea that telescopic phenomena are faithful images of the sky.

According to the Copernican theory, Mars and Venus approach and recede from the earth by a factor of 1:6 or 1:8, respectively. (These are approximate numbers.) Their change of brightness should be 1:40 and 1:60, respectively (these are Galileo's values). Yet Mars changes very little and the variation in the brightness of Venus 'is almost imperceptible'.[1] These experiences 'overtly contradict the annual movement [of the earth]'.[2] The telescope, on the other hand, produces new and strange *phenomena*, some of them exposable as illusory by observation with the naked eye, some contradictory, some having even the appearance of being illusory, while the only *theory* that could have brought order into this chaos, Kepler's theory of vision, is refuted by evidence of the plainest kind possible. But – and with this I come to what I think is a central feature of Galileo's procedure – *there are telescopic phenomena*, namely the telescopic variation of the brightness of the planets, *which agree more closely with Copernicus than do the results of naked-eye observation*. Seen through the telescope, Mars does indeed change as it should according to the Copernican view. Compared with the total performance of the telescope this change is still quite puzzling. It is just as puzzling as is the Copernican theory when compared with the pre-telescopic evidence. But the change is in harmony with the predictions of Copernicus. *It is this harmony* rather than any deep understanding of cosmology and of optics *which for Galileo proves Copernicus and the veracity of the telescope* in terrestrial *as well as* celestial matters. And it is this harmony on which he builds an entirely new view of the universe. 'Galileo,' writes Ludovico

---

1. The actual variations of Mars and Venus are four magnitudes and one magnitude respectively.
2. *Dialogue*, op. cit., p. 328.

Geymonat,[3] referring to this aspect of the situation, 'was not the first to turn the telescope upon the heavens, but . . . he was the first to grasp the enormous interest of the things thus seen. And he understood at once that these things fitted in perfectly with the Copernican theory whereas they contradicted the old astronomy. Galileo had believed for years in the truth of Copernicanism, but he had never been able to demonstrate it despite his exceedingly optimistic statements to friends and colleagues [he had not even been able to remove the refuting instances, as we have seen, and as he says himself]. Should direct proof [should even mere *agreement* with the evidence] be at last sought here? The more this conviction took root in his mind, the clearer to him became the importance of the new instrument. In Galileo's own mind faith in the reliability of the telescope and recognition of its importance were not *two separate acts*, rather, they were *two aspects of the same process*.' Can the absence of independent evidence be expressed more clearly? 'The *Nuncius*', writes Franz Hammer in the most concise account I have read of the matter,[4] 'contains two unknowns, the one being solved with the help of the other.' This is entirely correct, except that the 'unknowns' were not so much unknown as known to be false, as Galileo on occasions says himself. It is this rather peculiar situation, this harmony between two interesting but refuted ideas, which Galileo exploits in order to prevent the elimination of either.

Exactly the same procedure is used to preserve his new dynamics. We have seen that this science, too, was endangered by observable events. To eliminate the danger Galileo introduces friction and other disturbances with the help of *ad hoc* hypotheses, treating them as tendencies *defined* by the obvious discrepancy between fact and theory rather than as physical events *explained* by a theory of friction for which new and independent evidence might some day become available (such a theory arose only much later, in the 18th century). Yet the agreement between

---

3. Op. cit., pp. 38ff (my italics).

4. *Johannes Kepler, Gesammelte Werke*, op. cit., Vol. IV, p. 447. Kepler (*Conversation*, op. cit., p. 14) speaks of 'mutually self-supporting evidence'. Remember, however, that what is 'mutually self-supporting' are two refuted hypotheses and *not* two hypotheses which have *independent support* in the domain of basic statements. In a letter to Herwarth of 26 March 1598, Kepler speaks of the 'many reasons' he wants to adduce for the motion of the earth, adding that 'each of these reasons, taken for itself, would find only scant belief (Caspar–Dyck, *Johannes Kepler in seinen Briefen*, Vol. 1, Munich, 1930, p. 68).

the new dynamics and the idea of the motion of the earth, which Galileo increases with the help of his method of *anamnesis*, makes both seem more reasonable.

The reader will realize that a more detailed study of historical phenomena such as these creates considerable difficulties for the view that the transition from the pre-Copernican cosmology to that of the 17th century consisted in the replacement of refuted theories by more general conjectures which explained the refuting instances, made new predictions, and were corroborated by observations carried out to test these new predictions. And he will perhaps see the merits of a different view which asserts that, while the pre-Copernican astronomy *was in trouble* (was confronted by a series of refuting instances and implausibilities), the Copernican theory *was in even greater trouble* (was confronted by even more drastic refuting instances and implausibilities); but that being in harmony *with still further inadequate theories* it gained strength, and was retained, the refutations being made ineffective by *ad hoc* hypotheses and clever techniques of persuasion. This would seem to be a much more adequate description of the developments at the time of Galileo than is offered by almost all alternative accounts.

I shall now interrupt the historical narrative to show that the description is not only *factually adequate*, but that it is also *perfectly reasonable*, and that any attempt to enforce some of the more familiar methodologies of the 20th century would have had disastrous consequences.

11

Such 'irrational' methods of support are needed because of the 'uneven development' (Marx, Lenin) of different parts of science. Copernicanism and other essential ingredients of modern science survived only because reason was frequently overruled in their past.

A prevalent tendency in philosophical discussions is to approach problems of knowledge *sub specie aeternitatis*, as it were. Statements are compared with each other without regard to their history and without considering that they might belong to different historical strata. For example, one asks: given background knowledge, initial conditions, basic principles, accepted observations – what conclusions can we draw about a newly suggested hypothesis? The answers vary considerably. Some say that it is possible to determine degrees of confirmation and that the hypothesis can be evaluated with their help. Others reject any logic of confirmation and judge hypotheses by their content, and by the falsifications that have actually occurred. But almost everyone takes it for granted that precise observations, clear principles and well-confirmed theories *are already decisive*; that they can and must be used *here and now* to either eliminate the suggested hypothesis, or to make it acceptable, or perhaps even to prove it.

Such a procedure makes sense only if we can assume that the elements of our knowledge – the theories, the observations, the principles of our arguments – are *timeless entities* which share the same degree of perfection, are all equally accessible, and are related to each other in a way that is independent of the events that produced them. This is, of course, an extremely common assumption. It is taken for granted by most logicians; it underlies the familiar distinction between a context of discovery and a context of justification; and it is often expressed by saying that science deals with propositions and not with statements or sentences. However, the procedure overlooks that science is a complex and heterogeneous *historical process* which contains vague and incoherent anticipations of future ideologies side by side with highly sophisticated theoretical systems and ancient and petrified forms of thought. Some of its elements are available in the form of neatly written statements while others are submerged and become known only by contrast, by comparison with new and unusual views. (This is the way in which the inverted tower argument helped Galileo to discover the natural interpretations hostile to

Copernicus. And this is also the way in which Einstein discovered certain deep-lying assumptions of classical mechanics, such as the assumption of the existence of infinitely fast signals. For general considerations, see the last paragraph of Chapter 5.) Many of the conflicts and contradictions which occur in science are due to this heterogeneity of the material, to this 'unevenness' of the historical development, as a Marxist would say, and they have no immediate theoretical significance.[1] They have much in common with the problems which arise when a power station is needed right next to a Gothic cathedral. Occasionally, such features are taken into account; for example, when it is asserted that physical laws (statements)

---

1. According to Marx, 'secondary' parts of the social process, such as demand, artistic production or legal relations, may get ahead of material production and drag it along: see *The Poverty of Philosophy* but especially the *Introduction to the Critique of Political Economy*, Chicago, 1918, p. 309: 'The unequal relation between the development of material production and art, for instance. In general, the conception of progress is not to be taken in the sense of the usual abstraction. In the case of art, etc., it is not so important and difficult to understand this disproportion as in that of practical social relations, e.g. the relation between education in the U.S. and Europe. The really difficult point, however, that is to be discussed here is that of the unequal development of relations of production as legal relations.' Trotsky describes the same situation: 'The gist of the matter lies in this, that the different aspects of the historical progress – economics, politics, the state, the growth of the working class – do not develop simultaneously along parallel lines' ('The School of Revolutionary Strategy', speech delivered at the general party membership meeting of the Moscow Organization of July 1921, published in *The First Five Years of the Communist International*, Vol. II, New York, 1953, p. 5). See also Lenin, *Left-Wing Communism – an Infantile Disorder* (op. cit., p. 59), concerning the fact that multiple causes of an event may be out of phase and have an effect only when they occur together. In a different form, the thesis of 'uneven development' deals with the fact that capitalism has reached different stages in different countries, and even in different parts of the same country. This second type of uneven development may lead to inverse relations between the accompanying ideologies, so that efficiency in production and radical political ideas develop in inverse proportions. 'In civilized Europe, with its highly developed machine industry, its rich, multiform culture and its constitutions, a point of history has been reached when the commanding bourgeoisie, fearing the growth and increasing strength of the proletariat, comes out in support of everything backward, moribund, and medieval. . . . But all young Asia grows a mighty democratic movement, spreading and gaining in strength' (Lenin, 'Backward Europe and Advanced Asia', *Collected Works*, Vol. 19, op. cit., pp. 99ff). For this very interesting situation, which deserves to be exploited for the philosophy of science, see A.C. Meyer, *Leninism*, Cambridge, 1957, Chapter 12 and L. Althusser, *For Marx*, London and New York, 1970, Chapters 3 and 6. The philosophical background is splendidly explained in Mao Tse-tung's essay *On Contradiction* (*Selected Readings*, Peking, 1970, p. 70, especially section IV).

and biological laws (statements) belong to different conceptual domains and cannot be directly compared. But in most cases, and especially in the case observation vs theory, our methodologies project the various elements of science and the different historical strata they occupy on to one and the same plane, and proceed at once to render comparative judgements. This is like arranging a fight between an infant and a grown man, and announcing triumphantly, what is obvious anyway, that the man is going to win (the history of science is full of inane criticisms of this kind and so is the history of psychoanalysis and of Marxism). In our examination of new hypotheses we must obviously take the historical situation into account. Let us see how this is going to affect our judgement!

The geocentric hypothesis and Aristotle's theory of knowledge and perception are well adapted to each other. Perception supports the theory of locomotion that entails the unmoved earth and it is in turn a special case of a comprehensive view of motion that includes locomotion, increase and decrease, qualitative alteration, generation and corruption. This comprehensive view defines motion as the transition of a form from an agent to a patient which terminates when the patient possesses exactly the same form that characterized the agent at the beginning of the interaction. Perception, accordingly, is a process in which the form of the object perceived enters the percipient as precisely the same form that characterized the object so that the percipient, in a sense, assumes the properties of the object.

A theory of perception of this kind (which one might regard as a sophisticated version of naive realism) does not permit any major discrepancy between observations and the things observed. 'That there should be things in the world which are inaccessible to man not only now, and for the time being, but in principle, and because of his natural endowment, and which would therefore never be seen by him – this was quite inconceivable for later antiquity as well as for the Middle Ages.'[2] Nor

---

2. F. Blumenberg, *Galileo Galilei, Sidereus Nuncius, Nachricht von neuen Sternen*, Vol. 1, Frankfurt, 1965, p. 13. Aristotle himself was more open-minded: 'The evidence (concerning celestial phenomena) is furnished but scantily by sensations, whereas respecting perishable plants and animals we have abundant information, living as we do in their midst . . .', *DePart. Anim.*, 644b26ff. In what follows, a highly idealized account is given of later Aristotelianism. Unless otherwise stated, the word 'Aristotle' refers to this idealization. For the difficulties in forming a coherent picture of Aristotle *himself* see Düring, *Aristoteles*, Heidelberg, 1966. For some differences between Aristotle and his mediaeval followers see Wolfgang Wieland, *Die Aristotelische Physik*, Göttingen, 1970.

does the theory encourage the use of instruments, for they interfere with the processes in the medium. These processes carry a true picture only as long as they are left undisturbed. Disturbances create forms which are no longer identical with the shape of the objects perceived – they create *illusions*. Such illusions can be readily demonstrated by examining the images produced by curved mirrors,[3] or by crude lenses (and remember that the lenses used by Galileo were far from the level of perfection achieved today): they are distorted, the lens-images have coloured fringes, they may appear at a place different from the place of the object and so on. Astronomy, physics, psychology, epistemology – all these disciplines collaborate with the Aristotelian philosophy to create a system that is coherent, rational and in agreement with the results of observation as can be seen from an examination of Aristotelian philosophy in the form in which it was developed by some mediaeval philosophers. Such an analysis shows the inherent power of the Aristotelian system.

The role of observation in Aristotle is quite interesting. Aristotle is an empiricist. His injunctions against an overly-theoretical approach are as militant as those of the 'scientific' empiricists of the 17th and 18th centuries. But while the latter take both the truth and the content of empiricism for granted, Aristotle explains the nature of experience and why it is important. Experience is what a normal observer (an observer whose senses are in good order and who is not drunk or sleepy, etc.) perceives under normal circumstances (broad daylight; no interference with the medium) and describes in an idiom that fits the facts and can be understood by all. Experience is *important for knowledge* because, given normal circumstances, the perceptions of the observer contain identically the same forms that reside in the object. Nor are these explanations *ad hoc*. They are a direct consequence of Aristotle's general theory of motion, taken in conjunction with the physiological idea that sensations obey the same physical laws as does the rest of the universe. And they are confirmed by the evidence that confirms either of these two views (the existence of distorted lens-images being part of the evidence). We understand today a little better why a theory of motion and perception

---

3. Already a plain mirror gives rise to an interesting illusion. To notice it, first look at yourself in a plain mirror. You will see your face at its 'normal' size. Then let some steam condense on the surface of the mirror and draw the outline of your face in the steam. The outline will look about half the size of your face.

which is now regarded as false could be so successful (evolutionary explanation of the adaptation of organisms; movement in media). The fact remains that no decisive empirical argument could be raised against it (though it was not free from difficulties).

FIGURE 2. Moon, age seven days (first quarter).

This harmony between human perception and the Aristotelian cosmology is regarded as illusory by the supporters of the motion of the earth. In the view of the Copernicans there exist large-scale processes which involve vast cosmic masses and yet *leave no trace* in our experience. The existent observations therefore count no longer as tests of the new basic laws that are being proposed. They are not directly attached to these laws, and they may be entirely disconnected. *Today, after* the success of modern science led to the belief that the relation between man and the universe is not as simple as is assumed by naive realism, we can say that this was a correct guess, that the observer is indeed separated from the laws of the world by the special physical conditions of his observation platform, the moving earth (gravitational effects; law of inertia; Coriolis forces; influence of the atmosphere upon optical observations; aberration; stellar parallax; and so on . . .); by the idiosyncrasies of his basic instrument of observation, the human eye (irradiation; after-images; mutual inhibition of adjacent retinal elements; and so on . . .); as well as by older views which have invaded the observation language and made it speak the language of naive realism (natural interpretations). Observations may contain a contribution from the thing observed, but this contribution merges with other effects (some of which we have just mentioned), and it may be completely obliterated by them. Just consider the image of a fixed star as viewed through a telescope. This image is displaced by the effects of refraction, aberration and, possibly, of gravitation. It contains the spectrum of the star not as it is now, but as it was some time ago (in the case of extra-galactic supernovae the difference may be millions of years), and distorted by Doppler effect, intervening galactic matter, etc. Moreover, the extension and the internal structure of the image is entirely determined by the telescope and the eyes of the observer: it is the telescope that decides how large the diffraction disks are going to be, and it is the human eye that decides how much of the structure of these disks is going to be seen. It needs considerable skill *and much theory* to isolate the contribution of the original cause, the star, and to use it for a test, but this means that non-Aristotelian cosmologies can be tested only after we have *separated* observations and laws with the help of auxiliary sciences describing the complex processes that occur between the eye and the object, and the even more complex processes between the cornea and the brain. We must *subdivide* what we perceive to find a core that mirrors the stimulus and nothing else. In the case of Copernicus we need a new *meteorology* (in the good old sense of the word, as dealing with things below the moon), a new science of *physiological optics* that deals with the subjective (mind) and the objective (light, medium, lenses, structure of the eye) aspects of vision, as well as a new *dynamics* stating the manner in which the motion of the earth might influence the physical processes at its surface. Observations become relevant only *after* the processes described by these new subjects have been inserted between the world and the eye. The language in which we express our observations may have to be revised as well so that the new cosmology receives a fair chance and is not endangered by an unnoticed collaboration of sensations and older ideas. In sum: *what is needed for a test of Copernicus is an entirely new world-view containing a new view of man and of his capacities of knowing.*[4]

---

4. Bacon realized that scientific change involves a reformation not only of a few ideas, but of an entire world-view and, perhaps, of the very nature of humans. 'For the senses are weak and erring', he writes in *Novum Organum*, Aphorism 50. 'For man's sense is falsely asserted to be the standard of things; on the contrary, all the perceptions, both of the senses and of the mind bear reference to man and not to the universe, and the human mind resembles those uneven mirrors which impart their own properties to different objects from which rays are emitted and distort and disfigure them' (Aphorism 41). Bacon repeatedly comments on the 'dullness, incompetency and errors of the senses' (50) and permits them only to 'judge . . . the experiment' while it is the experiment that functions as a judge 'of nature and the thing itself' (50). Thus when Bacon speaks of the 'unprejudiced senses' he does not mean sense-data, or immediate impressions, but reactions of a sense organ *that has been rebuilt* in order to mirror nature in the right way. Research demands *that the entire human being be rebuilt.* This idea of a physical and mental reform of humanity has religious features. A 'demolishing branch' (115), an 'expiatory process', a 'purification of the mind' (69) must precede the accumulation of knowledge. 'Our only hope of salvation is to begin the whole labour of the mind again' (Preface) but only 'after having cleansed, polished, and levelled its surface' (115). Preconceived notions (36), opinions (42ff), even the most common words (59, 121) 'must be abjured and renounced with firm and solemn resolution . . . so that the access to the kingdom of man, which is founded on the sciences, may resemble that to the kingdom of heaven, where no admission is conceded except to children' (68).

A reform of man is necessary for a correct science – but it is not sufficient. Science, according to Bacon, not only orders events, it is also supposed to give physical reasons. Thus Ptolemy and Copernicus give us 'the number, situation, motion, and periods of the stars, as a beautiful outside of the heavens, whilst the flesh and the entrails are wanting; that is, a well fabricated system, or the physical reasons and foundations for a just theory, that should not only solve phenomena, as almost any ingenious theory may do, but show the substance, motions and influences of the heavenly bodies as they really are.' *Advancement of Learning*, Chapter 4, quoted from Wiley Books, New York, 1944, p. 85. See also the *Novum Organum*, op. cit., p. 371: 'For let no one hope to determine the question whether the earth or heaven revolve in the diurnal motion, unless he have first comprehended the nature of spontaneous motion': the new man needs a new physics in order to give substance to his astronomy. Galileo did not succeed in providing such a physics.

Science-loving philosophers, including those who call themselves 'critical', are quick to criticize thinkers who do not share their pet ideas. Bacon was often criticized for not at once falling for Copernicus. He was criticized for this unspeakable crime by philosophers whose own 'rationalism' would never have allowed Copernicus to live. An example is K.R. Popper, *The Open Society and Its Enemies*, Vol. 2, p. 16.

It is obvious that such a new world-view will take a long time appearing, and that we may never succeed to formulate it in its entirety. It is extremely unlikely that the idea of the motion of the earth will at once be followed by the arrival, in full formal splendour, of all the sciences that are now said to constitute the body of 'classical physics'. Or, to be a little more realistic, such a sequence of events is not only extremely unlikely, *it is impossible in principle*, given the nature of humans and the complexities of the world they inhabit. Today Copernicus, tomorrow Helmholtz – this is but a Utopian dream. Yet it is only *after* these sciences have arrived that a test can be said to make sense.

This need to *wait*, and to *ignore* large masses of critical observations and measurements, is hardly ever discussed in our methodologies. Disregarding the possibility that a new physics or a new astronomy might have to be judged by a new theory of knowledge and might require entirely new tests, empirically inclined scientists at once confront it with the *status quo* and announce triumphantly that 'it is not in agreement with facts and received principles'. They are of course right, and even trivially so, but not in the sense intended by them. For at an early stage of development the contradiction only indicates that the old and the new are *different* and *out of phase*. It does not show which view is the *better* one. A judgement of *this* kind presupposes that the competitors confront each other on equal terms. How shall we proceed in order to bring about such a fair comparison?

The first step is clear: we must *retain* the new cosmology until it has been supplemented by the necessary auxiliary sciences. We must retain it in the face of plain and unambiguous refuting facts. We may, of course, try to explain our action by saying that the critical observations are either not relevant or that they are illusory, but we cannot support such an explanation by a single objective reason. Whatever explanation we give is nothing but a *verbal gesture*, a gentle invitation to participate in the development of the new philosophy. Nor can we reasonably remove the received *theory* of perception which says that the observations are relevant, gives reasons for this assertion, and is confirmed by independent evidence. Thus the new view is arbitrarily separated from data that supported its predecessor and is made more 'metaphysical': a new period in the history of science commences with a *backward movement* that returns us to an earlier stage where theories were more vague and had smaller empirical content. This backward movement is not just an accident; it has a definite function; it is essential if we want to overtake the *status quo*, for it gives us the time and the freedom that are needed for developing the main view in detail, and for finding the necessary auxiliary sciences.[5]

This backward movement is indeed essential – but how can we persuade people to follow our lead? How can we lure them away from a well-defined, sophisticated and empirically successful system and make them transfer their allegiance to an unfinished and absurd hypothesis? To a hypothesis, moreover, that is contradicted by one observation after another if we only take the trouble to compare it with what is plainly shown to be the case by our senses? How can we convince them that the success of the *status quo* is only apparent and is bound to be shown as such in 500 years or more, when there is not a single argument on our side (and remember that the illustrations I used two paragraphs earlier derive their force from the successes of classical physics and were not available to the Copernicans).[6] It is clear that allegiance to the new ideas will have to be brought about by means other than arguments. It will have to be brought about by *irrational means* such as propaganda, emotion, *ad hoc* hypotheses, and appeal to prejudices of all kinds. We need these 'irrational means' in order to uphold what is nothing but a blind faith until we have found the auxiliary sciences, the facts, the arguments that turn the faith into sound 'knowledge'.

It is in this context that the rise of a new secular class with a new outlook and considerable contempt for the science of the schools, its methods, its results, even for its language, becomes so important. The barbaric Latin spoken by the scholars, the intellectual squalor of academic science, its other-worldliness which is soon interpreted as uselessness, its connection with the Church – all these elements are now

---

5. An example of a backward movement of this kind is Galileo's return to the kinematics of the *Commentariolus* and his disregard for the machinery of epicycles as developed in the *De Revol.* For an admirable *rational* account of this step see Imre Lakatos and Eli Zahar, 'Why Did Copernicus' Research Programme Supersede Ptolemy's?', in *Imre Lakatos, Philosophical Papers*, Vol. I, Cambridge, 1978.

6. They were available to the sceptics, especially to Aenesidemus, who points out, following Philo, that no object appears as it is but is modified by being combined with air, light, humidity, heat, etc.; see *Diogenes Laertius*, IX, 84. However, it seems that the sceptical view had only little influence on the development of modern astronomy, and understandably so: one does not start a movement by being reasonable.

lumped together with the Aristotelian cosmology, and the contempt one feels for them is transferred to every single Aristotelian argument.[7] This guilt-by-association does not make the arguments less *rational*, or less conclusive, *but it reduces their influence* on the minds of those who are willing to follow Copernicus. For Copernicus now stands for progress in other areas as well, he is a symbol for the ideals of a new class that looks back to the classical times of Plato and Cicero and forward to a free and pluralistic society. The association of astronomical ideas and historical and class tendencies does not produce new arguments either. But it engenders a firm commitment to the heliocentric view – and this is all that is needed at this stage, as we have seen. We have also seen how masterfully Galileo exploits the situation and how he amplifies it by tricks, jokes, and *nonsequiturs* of his own.[8]

---

7. For these social pressures see Olschki's magnificent *Geschichte der neusprachlichen wissenschaftlichen Literatur*. For the role of Puritanism see R.F. Jones, op. cit., Chapters V and VI.

8. In a remarkable book, *Galileo: Heretic*, Princeton, 1987 (first published, in Italian, in 1982), Pietro Redondi has described the groups both inside the Church (and including the Pope himself) and outside of it who looked favourably upon new scientific developments – the views on perception, continuity, matter and motion that had been explained by Galileo in his *Assayer* among them. Being in direct conflict with the traditional account of the Eucharist, the most important sacrament, these views were considerably more dangerous than Copernicanism and could be tolerated only as long as the groups and the Pope himself had the upper hand in the complex political developments of the time (Thirty Years' War; French and Spanish politics; the French alliance with the Pope). The political reversal of the Pope's fortunes, the accusations of leniency towards heretics that were raised against him on political grounds, cast a shadow on his attitude towards scientific matters as well (here, too, he seemed to support heresy) and made protective measures necessary. Redondi tries to show (a) that the physics of the time was connected with theological doctrines such as the doctrine of the Eucharist and that a history of science that neglects the connection becomes incomprehensible and (b) that the attitude towards scientific problems caused by the connection and thus the attitude towards innovation changed with the political climate. The second part of (b) may well be true but there is only weak evidence to support the rest: what Galileo says about atomism in the *Assayer* is much too brief and indefinite to conflict with transsubstantiation (it is an aside almost, not an elaborate statement) and with the exception of a rather problematic document no such conflict was perceived. (See R.S. Westfall, *Essays on the Trial of Galileo*, Vatican Observatory Publications, 1989, pp. 84ff.) What is valuable in Redondi's account is that he widens the domain of possible influences and thus undermines the (anachronistic) belief that then as now scientific rationality was restricted to the internal problem situation of a scientific discipline.

We are here dealing with a situation that must be analysed and understood if we want to adopt a more reasonable attitude towards the issue between 'reason' and 'irrationality'. Reason grants that the ideas which we introduce in order to expand and to improve our knowledge may *arise* in a very disorderly way and that the *origin* of a particular point of view may depend on class prejudice, passion, personal idiosyncrasies, questions of style, and even on error, pure and simple. But it also demands that in *judging* such ideas we follow certain well-defined rules: our *evaluation* of ideas must not be invaded by irrational elements. Now, what our historical examples seem to show is this: there are situations when our most liberal judgements and our most liberal rules would have eliminated a point of view which we regard today as essential for science, and would not have permitted it to prevail – and such situations occur quite frequently. The ideas survived and they *now* are said to be in agreement with reason. They survived because prejudice, passion, conceit, errors, sheer pigheadedness, in short because all the elements that characterize the context of discovery, *opposed* the dictates of reason *and because these irrational elements were permitted to have their way.* To express it differently: *Copernicanism and other 'rational' views exist today only because reason was overruled at some time in their past.* (The opposite is also true: witchcraft and other 'irrational' views have *ceased* to be influential only because reason was overruled at some time in *their* past.)⁹

Now, assuming that Copernicanism is a Good Thing, we must also admit that its survival is a Good Thing. And, considering the conditions of its survival, we must further admit that it was a Good Thing that reason was overruled in the 16th, 17th and even the 18th centuries. Moreover, the cosmologists of the 16th and 17th centuries did not have the knowledge we have today, they did not know that Copernicanism was capable of giving rise to a scientific system that is acceptable from

---

9. These considerations refute J. Dorling, who, in *British Journal for the Philosophy of Science*, Vol. 23, 1972, p. 189f, presents my 'irrationalism' as a presupposition of my research, not as a result. He continues: '. . . one would have thought that the philosopher of science would be most interested in picking out and analysing in detail those scientific arguments which did seem to be rationally reconstructible.' One would have thought that the philosopher of science would be most interested in picking out and analysing in detail those moves which are necessary for the *advancement* of science. Such moves, I have tried to show, often resist rational reconstruction.

the point of view of 'scientific method'. They did not know which of the many views that existed at their time would lead to future reason when defended in an 'irrational' way. Being without such guidance they had to make a guess, and in making this guess they could only follow their inclinations, as we have seen. Hence it is advisable to let one's inclinations go against reason *in any circumstances*, for it makes life less constrained and science may profit from it.

It is clear that this argument, that advises us not to let reason overrule our inclinations and occasionally to suspend reason altogether, does not depend on the historical material which I have presented. If my account of Galileo is historically correct, then the argument stands as formulated. If it turns out to be a fairy-tale, then this fairy-tale tells us that a conflict between reason and the preconditions of progress is *possible*, it indicates how it might arise, and it forces us to conclude that our chances to progress *may* be obstructed by our desire to be rational. And note that progress is here defined as a rationalistic lover of science would define it, i.e. as entailing that Copernicus is better than Aristotle and Einstein better than Newton. Of course, there is no need to accept this definition, which is certainly quite narrow. I use it only to show that an idea of reason accepted by the majority of rationalists may prevent progress as defined by the very same majority. I now resume the discussion of some details of the transition from Aristotle to Copernicus.

The first step on the way to a new cosmology, I have said, is a step *back*: apparently relevant evidence is pushed aside, new data are brought in by *ad hoc* connections, the empirical content of science is drastically reduced. Now the cosmology that happens to be at the centre of attention and whose adoption causes us to carry out the changes just described differs from other views in one respect only: it has features which at the time in question seem attractive to some people. But there is hardly any idea that is totally without merit and that might not also become the starting point of concentrated effort. No invention is ever made in isolation, and no idea is, therefore, completely without (abstract or empirical) support. Now if partial support and partial plausibility suffice to start a new trend – and I have suggested that they do – if starting a new trend means taking a step back from the evidence, if any idea can become plausible and can receive partial support, then the step back is in fact a step forward, and away from the tyranny of tightly-knit, highly corroborated, and gracelessly presented theoretical systems. 'Another different error', writes Bacon on precisely this point,[10] 'is the . . . peremptory reduction of knowledge into arts and methods, from which time the sciences are seldom improved; for as young men rarely grow in stature after their shape and limbs are fully formed, so knowledge, whilst it lies in aphorisms and observations, remains in a growing state; but when once fashioned into methods, though it may be further polished, illustrated and fitted for use, is no longer increased in bulk and substance.'

The similarity with the arts which has often been asserted arises at exactly this point. Once it has been realized that a close empirical fit is no virtue and that it must be relaxed in times of change, then style, elegance of expression, simplicity of presentation, tension of plot and narrative, and seductiveness of content become important features of our knowledge. They give life to what is said and help us to overcome the resistance of the observational material.[11] They *create* and maintain interest in a theory that has been partly removed from the observational plane and would be inferior to its rivals when judged by the customary standards. It is in this context that much of Galileo's work should be seen. This work has often been likened to *propaganda*[12] – and propaganda it certainly is. But propaganda of this kind is not a marginal affair that surrounds allegedly more substantial means of defence, and that should perhaps be avoided by the 'professionally honest scientist'. In the circumstances we are considering now, *propaganda is of the essence*. It is of the essence because interest must be created at a time when the usual methodological prescriptions have no point of attack; and because this interest must be maintained, perhaps for centuries, until new reasons arrive. It is also clear that such reasons, i.e. the appropriate auxiliary sciences, need not at once turn up in full formal splendour. They may at first be quite inarticulate, and may even conflict with the existing evidence. Agreement, or partial agreement, with the cosmology is all that is needed in the beginning. The agreement shows that they are at least *relevant* and that they may some day produce full-fledged positive evidence. Thus the idea that the

---

10. *Advancement of Learning* (1605 edition), New York, 1944, p. 21. See also the *Novum Organum*, Aphorisms 79, 86, as well as J.W.N. Watkins' splendid little book *Hobbes' System of Ideas*, London, 1965, p. 169.

11. 'What restitutes to scientific phenomenon its life, is art' (*The Diary of Anaïs Nin*, Vol I, p. 277).

12. See A. Koyré, *Etudes Galiléennes*, Vol. III, Paris, 1939, pp. 53ff.

telescope shows the world as it really is leads to many difficulties. But the support it lends to, and receives from, Copernicus is a hint that we might be moving in the right direction.

We have here an extremely interesting relation between a general view and the particular hypotheses which constitute its evidence. It is often assumed that general views do not mean much unless the relevant evidence can be fully specified. Carnap, for example, asserts that 'there is no independent interpretation for [the language in terms of which a certain theory or world-view is formulated]. The system T [the axioms of the theory and the rules of derivation] is itself an uninterpreted postulate system. [Its] terms obtain only an indirect and incomplete interpretation by the fact that some of them are connected by correspondence rules with observational terms.'[13] 'There is no independent interpretation,' says Carnap; and yet an idea such as the idea of the motion of the earth, which was inconsistent with the contemporary evidence, which was upheld by declaring this evidence to be irrelevant and which was therefore cut from the most important facts of contemporary astronomy, managed to become a nucleus, a crystallization point for the aggregation of other inadequate views which gradually increased in articulation and finally fused into a new cosmology including new kinds of evidence. There is no better account of this process than the description which John Stuart Mill has left us of the vicissitudes of his education. Referring to the explanations which his father gave him on logical matters, he writes: 'The explanations did not make the matter at all clear to me at the time; but they were not therefore useless; they remained as a nucleus for my observations and reflections to crystallize upon; the import of his general remarks being interpreted to me, by the particular instances which came under my notice *afterwards*.'[14] In exactly the same manner the Copernican view, though devoid of cognitive content from the point of view of a strict empiricism or else refuted, was needed in the construction of the supplementary sciences even before it became testable with their help, and even before it, in turn, provided them with supporting evidence of the most forceful kind.

---

13. 'The Methodological Character of Theoretical Concepts', *Minnesota Studies in the Philosophy of Science*, Vol. I, Minneapolis, p. 47.

14. *Autobiography*, quoted from *Essential Works of John Stuart Mill*, ed. Lerner, New York, 1965, p. 21.

There is a further element in this tapestry of moves, influences, beliefs which is rather interesting and which received attention only recently – the role of patronage. Today most researchers gain a reputation, a salary and a pension by being associated with a university and/or a research laboratory. This involves certain conditions such as an ability to work in teams, a willingness to subordinate one's ideas to those of a team leader, a harmony between one's ways of doing science and those of the rest of the profession, a certain style, a way of presenting the evidence – and so on. Not everyone fits conditions such as these; able people remain unemployed because they fail to satisfy some of them. Conversely the reputation of a university or a research laboratory rises with the reputation of its members. In Galileo's time patronage played a similar role. There were certain ways of gaining a patron and of keeping him. The patron in turn rose in estimation only if he succeeded to attract and to keep individuals of outstanding achievement. According to Westfall,[15] the Church permitted the publication of Galileo's *Dialogue* in the full knowledge of the controversial matters contained in it '[n]ot least because a Pope [Urban VIII] who gloried in his reputation as a Maecenas, was unwilling to place it in jeopardy by saying no to the light of his times', and Galileo fell because he violated his side of the rules of patronage.[16]

Considering all these elements, the 'Rise of the Copernican World-View' becomes a complicated matter indeed. Accepted methodological rules are put aside because of social requirements (patrons need to be persuaded by means more effective than argument), instruments are used to redefine experience instead of being tested by it, local results are extrapolated into space despite reasons to the contrary, analogies abound – and yet all this turns out, in retrospect, to have been the correct way of circumventing the restrictions implied by the human condition. This is the material that should be used to get better insight into the complex process of knowledge acquisition and improvement.

To sum up the content of the last five chapters:

---

15. Op. cit., p. 73.

16. Further details on these matters in Chapter 8, footnote 12 of the present essay, Westfall, op. cit., and M. Biagioli, *Galileo, Courtier*. M. Finocchiaro, *Galileo and the Art of Reasoning*, Dordrecht, 1980, has commented on Galileo's use of rhetoric, while M. Pera and W.R. Shea (eds), *Persuading Science – The Art of Scientific Rhetoric*, 1991, and especially Marcello Pera, *Science and Rhetoric*, forthcoming, comment on scientific rhetoric in general.

When the 'Pythagorean idea' of the motion of the earth was revived by Copernicus it met with difficulties which exceeded the difficulties encountered by contemporary Ptolemaic astronomy. Strictly speaking, one had to regard it as refuted. Galileo, who was convinced of the truth of the Copernican view and who did not share the quite common, though by no means universal, belief in a stable experience, looked for new kinds of fact which might support Copernicus and still be acceptable to all. Such facts he obtained in two different ways. First, by the invention of his *telescope*, which changed the *sensory core* of everyday experience and replaced it by puzzling and unexplained phenomena; and by his *principle of relativity and his dynamics*, which changed its *conceptual components*. Neither the telescopic phenomena nor the new ideas of motion were acceptable to common sense (or to the Aristotelians). Besides, the associated theories could be easily shown to be false. Yet these false theories, these unacceptable phenomena, were transformed by Galileo and converted into strong support of Copernicus. The whole rich reservoir of the everyday experience and of the intuition of his readers is utilized in the argument, but the facts which they are invited to recall are arranged in a new way, approximations are made, known effects are omitted, different conceptual lines are drawn, so that a *new kind of experience* arises, *manufactured* almost out of thin air. This new experience is then *solidified* by insinuating that the reader has been familiar with it all the time. It is solidified and soon accepted as gospel truth, despite the fact that its conceptual components are vastly more speculative than are the conceptual components of common sense. Following positivistic usage we may therefore say that Galileo's science rests on an *illustrated metaphysics*. The distortion permits Galileo to advance, but it prevents almost everyone else from making his effort the basis of a critical philosophy (for a long time emphasis was put either on his mathematics, or on his alleged experiments, or on his frequent appeal to the 'truth', and his propagandistic moves were altogether neglected). I suggest that what Galileo did was to let refuted theories support each other, that he built in this way a new world-view which was only loosely (if at all!) connected with the preceding cosmology (everyday experience included), that he established fake connections with the perceptual elements of this cosmology which are only now being replaced by genuine theories (physiological optics, theory of continua), and that whenever possible he replaced old facts by a new type of experience which he simply *invented* for the purpose of supporting Copernicus. Remember, incidentally, that Galileo's procedure drastically reduces the content of dynamics: Aristotelian dynamics was a general theory of change comprising locomotion, qualitative change, generation and corruption. Galileo's dynamics and its successors deal with locomotion only, other kinds of motion being pushed aside with the promissory note (due to Democritos) that locomotion will eventually be capable of comprehending *all* motion. Thus, a comprehensive empirical theory of motion is replaced by a much narrower theory plus a metaphysics of motion, just as an 'empirical' experience is replaced by an experience that contains speculative elements. This, I suggest, was the actual procedure followed by Galileo. Proceeding in this way he exhibited a style, a sense of humour, an elasticity and elegance, and an awareness of the valuable weaknesses of human thinking, which has never been equalled in the history of science. Here is an almost inexhaustible source of material for methodological speculation and, much more importantly, for the recovery of those features of knowledge which not only inform, but which also delight us.[17]

---

17. A few years ago Martin Gardner, the pitbull of scientism, published an article with the title 'Anti-Science, the Strange Case of Paul Feyerabend', *Critical Inquiry*, Winter 1982/83. The valiant fighter seems to have overlooked these and other passages. I am not against science. I praise its foremost practitioners and (next chapter) suggest that their procedures be adopted by philosophers. What I object to is narrow-minded philosophical interference and a narrow-minded extension of the latest scientific fashions to all areas of human endeavour – in short what I object to is a rationalistic interpretation and defence of science.

12

Galileo's method works in other fields as well. For example, it can be used to eliminate the existing arguments against materialism and to put an end to the philosophical *mind/body problem. (The corresponding* scientific *problems remain untouched, however.) It does not follow that it should be universally applied.*

Galileo made progress by changing familiar connections between words and words (he introduced new concepts), words and impressions (he introduced new natural interpretations), by using new and unfamiliar principles such as his law of inertia and his principle of universal relativity, and by altering the sensory core of his observation statements. His motive was the wish to accommodate the Copernican point of view. Copernicanism clashes with some obvious facts, it is inconsistent with plausible, and apparently well-established, principles, and it does not fit in with the 'grammar' of a commonly spoken idiom. It does not fit in with the 'form of life' that contains these facts, principles, and grammatical rules. But neither the rules, nor the principles, nor even the facts are sacrosanct. The fault may lie with them and not with the idea that the earth moves. We may therefore change them, create new facts and new grammatical rules, and see what happens once these rules are available and have become familiar. Such an attempt may take considerable time, and in a sense the Galilean venture is not finished even today. But we can already see that the changes were wise ones to make and that it would have been foolish to stick with the Aristotelian form of life to the exclusion of everything else.

With the mind/body problem, the situation is exactly the same. We have again observations, concepts, general principles, and grammatical rules which, taken together, constitute a 'form of life' that apparently supports some views, such as dualism, and excludes others, such as materialism. (I say 'apparently' for the situation is much less clear here than it was in the astronomical case.) And we may again proceed in the Galilean manner, look for new natural interpretations, new facts, new grammatical rules, new principles which can accommodate materialism and then compare the *total* systems – materialism and the new facts, rules, natural interpretations, and principles on the one side; dualism and the old 'forms of life' on the other. Thus there is no need to try, like Smart, to show that materialism is compatible with the ideology of common sense. Nor is the suggested procedure as 'desperate' (Armstrong) as it must appear to those who are unfamiliar with conceptual change. The procedure was commonplace in antiquity and it occurs wherever imaginative researchers strike out in new directions (Einstein and Bohr are recent examples).[1]

So far the argument was purely intellectual. I tried to show that neither logic nor experience can limit speculation and that outstanding researchers often transgressed widely accepted limits. But concepts have not only a logical content; they also have associations, they give rise to emotions, they are connected with images. These associations, emotions and images are essential for the way in which we relate to our fellow human beings. Removing them or changing them in a fundamental way may perhaps make our concepts more 'objective', but it often violates important social constraints. It was for this reason that Aristotle refused to abandon an intuitive view of human beings simply because a more physiological approach showed successes in a limited domain. For him a person was a social entity and defined by his or her function in the city no matter what atomists or physicians involved in theory might say. Similarly the Roman Church, being interested in souls and not only in astronomical tricks, forbade Galileo to present his badly founded guesses as truths and punished him when he violated the prohibition. The trial of Galileo raises important questions about the role products of specialists, such as abstract knowledge, are supposed to play in society. It is for this reason that I shall now give a brief account of this event.

---

1. For a more detailed discussion the reader is referred to my *Philosophical Papers*, Vol. 1, Chapters 9 and 10.

13

The Church at the time of Galileo not only kept closer to reason as defined then and, in part, even now; it also considered the ethical and social consequences of Galileo's views. Its indictment of Galileo was rational and only opportunism and a lack of perspective can demand a revision.

There were many trials in the 17th century. The proceedings started either with accusations made by private parties, with an official act by a public officer, or with an inquiry based on sometimes rather vague suspicions. Depending on the location, the distribution of jurisdiction and the balance of power at a particular time, crimes might be examined by secular courts such as the courts of kings or of free cities, by Church courts, such as the spiritual courts attached to every episcopate, or by the special courts of the Inquisition. After the middle of the 12th century the episcopal courts were greatly aided by the study of Roman law. Lawyers became so influential that, even if wholly untrained in canon law and theology, they had a much better chance of high preferment than a theologian.[1] The inquisitorial process removed safeguards provided by Roman law and led to some well-publicized excesses. What has not been publicized to the same extent is that the excesses of royal or secular courts often matched those of the Inquisition. It was a harsh and cruel age.[2] By 1600 the Inquisition had

---

1. For this complaint (made by Roger Bacon) see H. Ch. Lea, *A History of the Inquisition of the Middle Ages*, Vol. I, p. 309. Chapters IXff explain the details of the inquisitorial procedure, the ways in which they differed from other procedures and the reasons for the difference. See also G.G. Coulton, *Inquisition and Liberty*, Boston, 1959, Chapters XI–XV.

2. Charles Henry Lea, the great liberal historian, writes: 'On the whole we may conclude that the secret prisons of the Inquisition were less intolerable places of abode than the episcopal and public gaols. The general policy respecting them was more humane and enlightened than that of other jurisdictions, whether in Spain or elsewhere, although negligent supervision allowed of abuses and there were ample resources of rigor in reserve, when the obstinacy of the impenitent was to be broken down.' *History of the Inquisition in Spain*, Vol. 2, New York, 1906, p. 534. Prisoners accused before secular courts occasionally committed crimes under the jurisdiction of the Church so that they might be handed over to the Inquisition: Henry Kamen, *Die Spanische Inquisition*, Munich, 1980, p. 17.

lost much of its power and aggressiveness. This was true especially in Italy, and more particularly in Venice.[3]

The courts of the Inquisition also examined and punished crimes concerning the production and the use of knowledge. This can be explained by their origin: they were supposed to exterminate *heresy*, i.e. complexes consisting of actions, assumptions and talk making people inclined towards certain beliefs. The surprised reader who asks what knowledge has to do with the law should remember the many legal, social and financial obstacles knowledge-claims face today. Galileo wanted his ideas to replace the existing cosmology, but he was forbidden to work towards that aim. Today the much more modest wish of creationists to have their views taught in schools side by side with other and competing views runs into laws setting up a separation of Church and State.[4] Increasing amounts of theoretical and engineering information are kept secret for military reasons and are thereby cut off from international exchange.[5] Commercial interests have the same restrictive tendency. Thus the discovery of superconductivity in ceramics at (relatively) high temperatures, which was the result of international collaboration, soon led to protective measures by the American government.[6] Financial arrangements can make or break a research programme and an entire profession. There are many

---

3. In 1356 the secular officials of Venice forbade the Inquisitor of Treviso to try his prisoners, seized his informants and tortured them on the charge of pilfering the property of the accused. Lea, *Inquisition in the Middle Ages*, Vol. II, p. 273.

4. A comprehensive report of one of the trials that resulted from the conflict has been published in *Science*, Vol. 215, 1982, pp. 934ff. Many other trials followed.

5. It seems that the need for secrecy in nuclear matters was first raised by the scientists themselves. See the report and the documents in Spencer R. Weart and Gertrude Weiss-Szilard (eds), *Leo Szilard, His Version of the Facts*, Cambridge, Mass., 1978, esp. Chapters 2ff. See also the material on the Oppenheimer case. The inventor of the telescope was forced to secrecy as the military importance of the contrivance was soon realized. See Chapter 8, footnote 24.

Research teams become very secretive when approaching what they think is a Big Discovery. After all, what is at stake are patents, consultancies in industry, money and, perhaps, the honour of a Nobel Prize. For a special case see R.M. Haze, *Superconductors*, London, 1988. The manipulation of knowledge by the courts is discussed, with many examples, by Peter W. Huber, *Galileo's Revenge*, New York, 1991.

6. *Science*, Vol. 237, 1987, pp. 476ff and 593f. An important step towards exclusiveness consisted in assigning part of the research to the military.

ways to silence people apart from forbidding them to speak – and all of them are being used today. The process of knowledge production and knowledge distribution was never the free, 'objective', and purely intellectual exchange rationalists make it out to be.

The trial of Galileo was one of many trials. It had no special features except perhaps that Galileo was treated rather mildly, despite his lies and attempts at deception.[7] But a small clique of intellectuals aided by scandal-hungry writers succeeded in blowing it up to enormous dimensions so that what was basically an altercation between an expert and an institution defending a wider view of things now looks almost like a battle between heaven and hell. This is childish and also very unfair towards the many other victims of 17th-century justice. It is especially unfair towards Giordano Bruno, who was burned but whom scientifically minded intellectuals prefer to forget. It is not a concern for humanity but rather party interests which play a major role in the Galileo hagiography. Let us therefore take a closer look at the matter.[8]

The so-called trial of Galileo consisted of two separate proceedings, or trials. The first occurred in 1616. The Copernican doctrine was examined and criticized. Galileo received an order, but he was not punished. The second trial took place in 1632/33. Here the Copernican doctrine was no longer the point at issue. Rather, what was considered was the question of whether Galileo had obeyed the order given him in the first trial, or

---

7. An example is Galileo's reply to the inquiries of 12 April 1633: Maurice A. Finocchiaro, *The Galileo Affair*, Berkeley and Los Angeles, 1989, p. 262, the first two lines. The reaction of an admirer is characteristic: 'This absurd pretence . . .' Geymonat, op. cit., p. 149.

8. It cannot be denied that pressure groups, personal grievances, envy, the fact that Galileo, 'being too infatuated with his own genius' was 'unsufferable' (Westfall, op. cit., pp. 52, 38) and the rules of patronage played an important role as they, or similar circumstances, do at every trial. However, the tensions between various groups of the Church on the one side and the demands for scientific autonomy on the other were real enough; after all, their modern successors (should the sciences be given the run of our educational institutions and of society as a whole or should they be treated like any other special interest group?) are still with us. Here the Church did the right thing: the sciences do *not* have the last word in humane matters, knowledge included.

The main documents pertaining to the trial were assembled and translated with comments and an introduction by Finocchiaro, op. cit. I shall use his translations. Accounts of the trials and their problems are found in G. de Santillana, *The Crime of Galileo*, Chicago, 1954, Geymonat, op. cit., Redondi, op. cit., and, most recently, in Westfall.

whether he had deceived the inquisitors into believing that the order had never been issued. The proceedings of both trials were published by Antonio Favaro in Vol. 19 of the National Edition of Galilean material. The suggestion, rather popular in the 19th century, that the proceedings contained falsified documents and that the second trial was therefore a farce, seems no longer acceptable.[9]

The first trial was preceded by denunciations and rumours, in which greed and envy played a part, as in many other trials. The Inquisition started to examine the matter. Experts (*qualificatores*) were ordered to give an opinion about two statements which contained a more or less correct account of the Copernican doctrine.[10] Their decision[11] concerned two points: what would today be called the *scientific content* of the doctrine and its *ethical (social) implications*.

On the first point the experts declared the doctrine to be 'foolish and absurd in philosophy' or, to use modern terms, they declared it to be unscientific. This judgement was made without reference to the faith, or to Church doctrine, but was based exclusively on the scientific situation of the time. It was shared by many outstanding scientists (Tycho Brahe having been one of them) – *and it was correct*[12] when based on the facts, the

---

9. One of the authors of the suggestion was the Galileo scholar Emil Wohlwill. His reasons, rather impressive at the time, are given in his *Der Inquisitionsprozess des Galileo Galilei*, Berlin, 1870. According to Wohlwill two documents of the proceedings, dated 25 and 26 February 1616 (Finocchiaro, op. cit., pp. 147f) are mutually contradictory. The first advises Galileo to treat Copernicus as a mathematical model; should he reject the advice, then he is forbidden to mention Copernicus in any form whatsoever. In the second document Galileo is advised as above and immediately forbidden (i.e. without waiting for his reaction) to mention Copernicus. Wohlwill thought the second document to be a forgery. This seems now refuted. See de Santillana, Chapter 13. Stillman Drake (appendix to Geymonat) devised a very intriguing hypothesis to explain the discrepancy.

10. Some critics used idiosyncrasies in the formulation as proof of a lack of comprehension on the part of the experts. But there was no need for the inquisitors to stick closely to the language of the authors they examined. Their account of Copernicanism was clear enough without such textual puritanism.

11. Finocchiaro, op. cit., p. 146.

12. Note that in rendering my judgements I rely on standards subscribed to by many modern scientists and philosophers of science. Returned to the early 17th century, these champions of rationality would have judged Galileo as the Aristotelians judged him then. Michelson, for example, would have been aghast at Galileo's attempt to get knowledge out of an instrument as little understood as the telescope and Rutherford, who was never too happy about the theory of relativity, would have produced one of his

theories and the standards of the time. Compared with those facts, theories and standards the idea of the motion of the earth was as absurd as were Velikovsky's ideas when compared with the facts, theories and standards of the fifties. A modern scientist really has no choice in this matter. He cannot cling to his own very strict standards and at the same time praise Galileo for defending Copernicus. He must either agree with the first part of the judgement of the Church experts, or admit that standards, facts and laws never decide a case and that an unfounded, opaque and incoherent doctrine can be presented as a fundamental truth. Only few admirers of Galileo have an inkling of this rather complex situation.

The situation becomes even more complex when we consider that the Copernicans changed not only views, but also standards for judging views. Aristotelians, in this respect not at all unlike modern epidemiologists, molecular biologists and 'empirical' sociologists who insist either on the examination of large statistical samples or on 'clearcut experimental steps' in Luria's sense, demanded strong empirical support while the Galileans were content with far-reaching, unsupported and partially refuted theories.[13] I do not criticize them for that; on the contrary, I favour Niels

---

characteristic rude remarks. Salvador Luria, an outstanding microbiologist who favours theories decidable by 'clear-cut experimental step[s]', would have relegated the debate to 'outfields of science' like 'sociology' and would have stayed away from it (*A Slot Machine, a Broken Test Tube*, New York, 1985, pp. 115, 119). For what Galileo suggested was no less than to regard as true a theory which had only analogies in its favour and which suffered from numerous difficulties. And he made this suggestion in public while even today it is a deadly sin for a scientist to address the public before having consulted his peers (example in A. Pickering, 'Constraints on Controversy: the Case of the Magnetic Monopole', *Social Studies of Science*, Vol. 11, 1981, pp. 63ff). All this is realized neither by 'progressive' (i.e. scientifically inclined) princes of the Church nor by scientists, so the discussion of the 'trial of Galileo' occurs in a dream world with only little relation to the real world we and Galileo inhabit. Further arguments on that point are found in Chapter 9 of *Farewell to Reason* and Chapter 19, below.

13. As indicated in Chapter 8, footnote 1, Galileo's law of inertia was in conflict with the Copernican as well as the Keplerian treatment of planetary motion. Galileo hoped for future accommodations. That was a sensible thing to do but not in agreement with some standards of his time and of today. Today a similar clash between theoreticians and empiricists occurs in the field of epidemiology. There are theoretical reasons to expect that X-rays and other forms of particulate radiation constitute a cancer-risk down to the smallest dose. Many epidemiologists demand empirical proof, however, though it is clear that events, when occurring below a certain threshold of frequency, cannot be detected in that way.

Bohr's 'this is not crazy enough'. I merely want to reveal the contradiction in the actions of those who praise Galileo and condemn the Church, but become as strict as the Church was at Galileo's time when turning to the work of their contemporaries.

On the second point, the social (ethical) implications, the experts declared the Copernican doctrine to be 'formally heretical'. This means it contradicted Holy Scripture as interpreted by the Church, and it did so in full awareness of the situation, not inadvertently (that would be 'material' heresy).

The second point rests on a series of assumptions, among them the assumption that Scripture is an important boundary condition of human existence and, therefore, of research. The assumption was shared by all great scientists, Copernicus, Kepler and Newton among them. According to Newton knowledge flows from two sources – the word of God – the Bible – and the works of God – Nature; and he postulated divine interventions in the planetary system, as we have seen.[14]

The Roman Church in addition claimed to possess the exclusive rights of exploring, interpreting and applying Holy Scripture. Lay people, according to the teaching of the Church, had neither the knowledge nor the authority to tamper with Scripture, and they were forbidden to do so. This comment, whose rigidity was a result of the new Tridentine Spirit,[15] should not surprise anyone familiar with the habits of powerful institutions. The attitude of the American Medical Association towards lay practitioners is as rigid as the attitude of the Church was towards lay interpreters – and it has the blessing of the law. Experts, or ignoramuses having acquired the formal insignia of expertise, always tried and often succeeded in securing for themselves exclusive rights in special domains. Any criticism of the rigidity of the Roman Church applies also to its modern scientific and science-connected successors.

Turning now from the form and the administrative backing of the objection to its content, we notice that it deals with a subject that is gaining increasing importance in our own times – the quality of human

---

14. Chapter 5, footnote 4. See also the literature in footnote 6 to Chapter 4. According to Galileo (letter to Grand Duchess Christina) the idea of the two sources goes back to Tertullian *adv. Marciones* (E. Evans, ed.), 1, 18.

15. For the exact wording see Denzinger–Schoenmetzer, *Enchiridion Symbolorum*, 36th edition, Freiburg, 1976, pp. 365f.

existence. Heresy, defined in a wide sense, meant a deviation from actions, attitudes and ideas that guarantee a well-rounded and sanctified life. Such a deviation might be, and occasionally was, encouraged by scientific research. Hence, it became necessary to examine the heretical implications of scientific developments.

Two ideas are contained in this attitude. First, it is assumed that the quality of life can be defined independently of science, that it may clash with demands which scientists regard as natural ingredients of their activity, and that science must be changed accordingly. Secondly, it is assumed that Holy Scripture as interpreted by the Holy Roman Church adumbrates a correct account of a well-rounded and sanctified life.

The second assumption can be rejected without denying that the Bible is vastly richer in lessons for humanity than anything that might ever come out of the sciences. Scientific results and the scientific ethos (if there is such a thing) are simply too thin a foundation for a life worth living. Many scientists agree with this judgement.[16]

---

16. Thus Konrad Lorenz, in his interesting if somewhat superficial book *Die Acht Todsünden der Zivilisierten Menschheit*, Munich, 1984 (first published in 1973), p. 70 writes: 'The erroneous belief that only what can be rationally grasped or even what can be proved in a scientific way constitutes the solid knowledge of mankind has disastrous consequences. It prompts the "scientifically enlightened" younger generation to discard the immense treasures of knowledge and wisdom that are contained in the traditions of every ancient culture and in the teachings of the great world religions. Whoever thinks that all this is without significance naturally succumbs to another, equally pernicious mistake, living in the conviction that science is able, as a matter of course, to create from nothing, and in a rational way, an entire culture with all its ingredients.' In a similar vein J. Needham, initiator and part-author of a great history of Chinese science and technology, speaks of 'scientific opium', meaning by it 'a blindness to the suffering of others'. *Time, the Refreshing River*, Nottingham, 1986.

'Rationalism', writes Peter Medawar (*Advice to a Young Scientist*, New York, 1979, p. 101), 'falls short of answering the many simple and childlike questions people like to ask; questions about origins and purposes such as are often contemptuously dismissed as non-questions, or pseudo-questions, although people understand them clearly enough and long to have an answer. These are intellectual pains that rationalists – like bad physicians confronted by ailments they cannot diagnose or cure – are apt to dismiss as "imagination".'

The clearest and most perceptive statement is found in Jacques Monod, *Chance and Necessity*, New York, 1972, p. 170 (text in brackets from p. 169): 'Cold and austere,' writes Monod, 'proposing no explanation but imposing an ascetic renunciation of all other spiritual fare, [the idea that objective knowledge is the only authentic source of truth] was not of a kind to allay anxiety but aggravated it instead. By a single stroke it claimed to sweep

They agree that the quality of life can be defined independently of science – which is the first part of the first assumption. At the time of Galileo there existed an institution – the Roman Church – watching over this quality in its own particular way. We must conclude that the second point – Copernicus being 'formally heretical' – was connected with ideas that are urgently needed today. The Church was on the right track.

But was it perhaps mistaken in rejecting scientific opinions inconsistent with its idea of a Good Life? In Chapter 3 I argued that knowledge needs a plurality of ideas, that well-established theories are never strong enough to terminate the existence of alternative approaches, and that a defence of such alternatives, being almost the only way of discovering the errors of highly respected and comprehensive points of view, is required even by a narrow philosophy such as empiricism. Now if it should turn out that it is also required on ethical grounds, then we have two reasons instead of one, rather than a conflict with 'science'.

Besides, the Church, and by this I mean its most outstanding spokesmen, was much more modest than that. It did not say: what contradicts the Bible as interpreted by us must go, no matter how strong the scientific reasons in its favour. A truth supported by scientific reasoning was not pushed aside. It was used to revise the interpretation of Bible passages apparently inconsistent with it. There are many Bible passages which seem to suggest a flat earth. Yet Church doctrine accepted the spherical earth as a matter of course. On the other hand the Church was not ready to change just because somebody had produced some vague guesses. It wanted *proof* – scientific proof in scientific matters. Here it acted no differently from modern scientific institutions: universities, schools and even research institutes in various countries usually wait a long time before they incorporate new ideas into their curricula. (Professor Stanley Goldberg has described the situation in the case of the special theory of relativity.) But there was as yet no convincing proof of the Copernican doctrine. Hence Galileo was advised to teach Copernicus *as a hypothesis*; he was forbidden to teach it *as a truth*.

---

away the traditions of hundreds of thousands of years, which had become one with human nature itself. It wrote an end to the ancient animist convenant between man and nature, leaving nothing in place of that precious bond but an anxious quest in a frozen universe of solitude. With nothing to recommend it but a certain puritan arrogance, how could such an idea win acceptance? It did not; it still has not. It has however commanded recognition; but that it did only because of its prodigious power of performance.'

This distinction has survived until today. But while the Church was prepared to admit that some theories might be true and even that Copernicus' might be true, given sufficient evidence,[17] there are now many scientists, especially in high energy physics, who view *all* theories as instruments of prediction and reject truth-talk as being metaphysical and speculative. Their reason is that the devices they use are so obviously designed for calculating purposes and that theoretical approaches so clearly depend on considerations of elegance and easy applicability that the generalization seems to make good sense. Besides, the formal properties of 'approximations' often differ from those of the basic principles, many theories are first steps towards a new point of view which at some future time may yield them as approximations, and a direct inference from theory to reality is therefore rather naive.[18] All this was known to 16th- and 17th-century scientists. Only a few astronomers thought of deferents and epicycles as real roads in the sky; most regarded them as roads on paper which might aid calculation but which had no counterpart in

---

17. In a widely discussed letter which Cardinal Roberto Bellarmino, master of controversial questions at the Collegio Romano, wrote on 12 April 1615 to Paolo Antonio Foscarini, a Carmelite monk from Naples who had inquired about the reality of the Copernican system, we find the following passage (Finocchiaro, op. cit., p. 68): '. . . if there were a true demonstration that the sun is at the center of the world and the earth in the third heaven, and that the sun does not circle the earth but the earth circles the sun, then one would have to proceed with great care in explaining the Scriptures that appear contrary, *and say rather that we do not understand them than that what is demonstrated is false.* But I will not believe that there is such a demonstration, until it is shown me. Nor is it the same to demonstrate that by supposing the sun to be at the center and the earth in heaven one can save the appearances, and to demonstrate that in truth the sun is at the center and the earth in heaven; for I believe the first demonstration may be available, but I have very great doubts about the second, and in case of doubt we must not abandon the Holy Scripture as interpreted by the Holy Fathers.' In his *Considerations on the Copernican Opinion*, Finocchiaro, op. cit., pp. 70ff, esp. pp. 85f, Galileo addresses precisely these points. He agrees that if the Copernican astronomers are 'not more than ninety percent right, they may be dismissed' but adds that 'if all that is produced by philosophers and astronomers on the opposite side is shown to be mostly false and wholly inconsequential, then the other side should not be disparaged, nor deemed paradoxical, so as to think that it could never be clearly proved': research should be permitted even if demonstrations are not yet available. This does not conflict with Bellarmino's suggestions; it did conflict and to a certain extent still does conflict with the attitude of many modern research institutions.

18. More on this point in Nancy Cartwright, *How the Laws of Physics Lie*, Oxford, 1983.

reality. The Copernican point of view was widely interpreted in the same way – as an interesting, novel and rather efficient model. The Church requested, both for scientific and for ethical reasons, that Galileo accept this interpretation. Considering the difficulties the model faced when regarded as a description of reality, we must admit that '[l]ogic was on the side of . . . Bellarmine and not on the side of Galileo,' as the historian of science and physical chemist Pierre Duhem wrote in an interesting essay.[19]

To sum up: the judgement of the Church experts was scientifically correct and had the right social intention, viz. to protect people from the machinations of specialists. It wanted to protect people from being corrupted by a narrow ideology that might work in restricted domains but was incapable of sustaining a harmonious life. A revision of the judgement might win the Church some friends among scientists, but would severely impair its function as a preserver of important human and superhuman values.[20]

---

19. *To Save the Phenomena*, Chicago, 1963, p. 78.

20. After some apparent willingness to consider the matter (see the address of Pope John Paul II on the centenary of Einstein's birth, published as an Epilogue in Paul Cardinal Poupard (ed.), *Galileo Galilei: Towards a Resolution of 350 Years of Debate*, Pittsburgh, 1987), Cardinal Joseph Ratzinger, who holds a position similar to that once held by Bellarmine, formulated the problem in a way that would make a revision of the judgement anachronistic and pointless. See his talk in Parma of 15 March 1990, partly reported in *Il Sabato*, 31 March 1990, pp. 80ff. As witnesses the Cardinal quoted Ernst Bloch (being merely a matter of convenience, the scientific choice between geocentrism and heliocentrism cannot overrule the practical and religious centricity of the earth), C.F. von Weizsäcker (Galileo leads directly to the atom bomb) and myself (the chapter heading of the present chapter). I commented on the speech in two interviews, *Il Sabato*, 12 May 1990, pp. 54ff and *La Repubblica*, 14 July 1990, p. 20.

Galileo's inquiries formed only a small part of the so-called Copernican Revolution. Adding the remaining elements makes it still more difficult to reconcile the development with familiar principles of theory evaluation.

Galileo was not the only scientist involved in the reform of physics, astronomy and cosmology. Neither did he deal with the whole area of astronomy. For example, he never studied the motion of the planets in as much detail as did Copernicus and Kepler and he probably never read the more technical parts of Copernicus' great work. That was not unusual. Then as now knowledge was subdivided into specialities; an expert in one field rarely was also an expert in another and distant field. And then as now scientists with widely diverging philosophies could and did comment on new suggestions and developments. Tycho Brahe was an outstanding astronomer; his observations contributed to the downfall of generally accepted views. He noticed the importance of Copernicus' cosmology – yet he retained the unmoved earth, on physical as well as on theological grounds. Copernicus was a faithful Christian and a good Aristotelian; he tried to restore centred circular motion to the prominence it once had, postulated a moving earth, rearranged the planetary orbits and gave absolute values for their diameters. The astronomers surrounding Melanchthon and his educational reform accepted and praised the first part of his achievement, but (with a single exception – Rheticus) either disregarded, or criticized, or reinterpreted (Osiander!) the second. And they often tried to transfer Copernicus' mathematical models to the Ptolemaic system.[1] Maestlin, Kepler's teacher, regarded comets as solid bodies and tried calculating the orbit of one of them. His (incorrect) result made him accept the Copernican arrangement of the planetary orbits (it still influenced Kepler). Maestlin respected Aristotle but regarded mathematical correctness and harmony as signs of physical truth. Galileo's approach had its own idiosyncrasies, it was more complex, more conjectural, partly adapted to the greater role theological considerations played in Italy, partly determined by the laws of rhetoric or patronage. Many different personalities, professions and groups guided by different beliefs and subjected to different constraints

---

1. Details and literature in R.S. Westman, 'The Melanchthon Circle, Rheticus, and the Wittenberg Interpretation of the Copernican Theory', *Isis*, Vol. 66, 1975, pp. 165ff.

contributed to the process that is now being described, somewhat summarily, as the 'Copernican Revolution'.

As I said at the beginning this process was not a simple thing but consisted of developments in a variety of subjects, among them the following: cosmology; physics; astronomy; the calculation of astronomical tables; optics; epistemology; and theology.

I draw these distinctions not 'in order to be precise' but because they reflect actually existing subdivisions of research. Physics, for example, was a general theory of motion that described change without reference to the circumstances under which it occurred. It comprised locomotion, the growth of plants and animals as well as the transition of knowledge from a wise teacher to an ignorant pupil. Aristotle's *Physics* and the many mediaeval commentaries on it give us an idea of the problems treated and the solutions proposed. Cosmology described the structure of the universe and the special motions that are found in it. A basic law of *physics* in the sense just explained was that a motion without motor comes to a standstill – the 'natural' situation of a body is rest (this includes lack of qualitative change). The 'natural' motions of *cosmology* were those that occurred without noticeable interference; examples are the upward motion of fire and the downward motion of stones. Aristotle's *On the Heavens* and the many mediaeval commentaries on it give us an idea of the problems and the views discussed in this domain.

The books I just mentioned were for advanced studies only. Introductory texts omitted problems and alternative suggestions, and concentrated on the bare bones of the ideas then held. One of the most popular introductory texts of cosmology, Sacrobosco's *de spera*, contained a sketch of the world, and described the main spheres without giving the details of their motions – the rest is silence.[2] Still, it was used as a basis for rather advanced critical comments down to Galileo's own time.

---

2. See Lynn Thorndyke (ed.), *The Sphere of Sacrobosco and Its Commentators*, Chicago, 1949. The elements and their motions are briefly mentioned in the first chapter, together with a simple argument in favour of the unmoved earth: the earth is situated in the centre (this is shown earlier by optical arguments, including the fact that the constellations have the same size, no matter where the daily rotation puts them) and 'quicquid a medio movetur versus circumferentiam ascendit. Terra a medio movetur, ergo ascendit, quod pro impossibile relinquitur' (p. 85). Equant, deferent and epicycle are mentioned in the fourth chapter together with the miraculous nature of the solar eclipse accompanying Christ's death.

Physics and cosmology claimed to make true statements. Theology, which also claimed to make true statements, was regarded as a boundary condition for research in these fields, though the strength of this requirement and of its institutional backing varied in time and with location. It was never a necessary boundary condition for astronomy, which dealt with the motions of the stars, but without claiming truth for its models. Astronomers were interested in models that might correspond to the actual arrangement of the planets, but they were not restricted to them. Handbooks of astronomy such as Ptolemy's handbook and the various popularizations based on it contained detailed astronomical models preceded by sketchy cosmological introductions. As far as these introductions were concerned, there existed only one cosmology – Aristotle's. Some of the handbooks also contained tables. Tables were a further step away from 'reality'. They not only used 'hypotheses', i.e. models that might not reflect the structure of reality, they also used approximations. But an astronomer's approximations did not always correspond to the excellence of his models. 'Advanced' (from our standpoint) models might be combined with crude approximations and thus give worse tables than their older counterparts.[3]

The separation between physics and cosmology on the one side and astronomy on the other was not only a practical fact; it also had a firm philosophical backing. According to Aristotle,[4] mathematics does not deal with real things but contains abstractions. There exists therefore an essential difference between physical subjects such as physics, cosmology, biology and psychology and mathematical subjects such as optics and astronomy. In the encyclopaedias of the early Middle Ages the separation was a matter of course. Optical textbooks only rarely dealt with astronomical matters.[5]

---

3. The example of Ptolemy–Copernicus is treated by Stanley E. Babb Jr. in *Isis*, Vol. 68, September 1977, especially p. 432.

4. *Met.*, Book xiii, Chapter 2; *Physics*, Book ii, Chapter 2. For an account and defence of Aristotle's theory of mathematics see Chapter 8 of my *Farewell to Reason*.

5. As an example I mention John Pecham's optics (quoted from David Lindberg (ed.), *John Pecham and the Science of Optics*, Madison, 1970): astronomical matters occur here on pp. 153 (moon illusion and the northward displacement of the sun and the fixed stars, explained by vapours near the horizon), 209 (scintillation of the stars explained by unevennesses of their rotating surfaces which reflect the sunlight), 218 (impossibility to determine the size of the stars from their appearances), 233 (stars appear to be smaller than they actually are), 225 (they are displaced towards the north at the horizon, and the more so, the greater their distance from the meridian).

Astronomy used basic optical laws such as the law of linear propagation, but the more complicated parts of optical theory were not well known. The same is true of epistemology. Galileo's arguments (and the arguments of Copernicus on which they are based) brought epistemology back into science (the same happened many years later, in connection with the quantum theory).

Now is it to be expected that a collection of relatively independent subjects, research strategies, arguments and opinions such as the one just mentioned will develop in a uniform way? Can we really assume that all the physicists, cosmologists, theologians and philosophers who reacted to the Copernican doctrine were guided by the same motives and reasons and that these reasons were not only accepted by them, but were also regarded as being binding for any scientist entering the scene? The ideas of an individual scientist such as Einstein may show a certain coherence,[6] and this coherence may be reflected in his standards and his theorizing. Coherence is to be expected in totalitarian surroundings that guide research either by laws, by peer pressure or by financial machinations. But the astronomers at the time of Copernicus and after did not live in such surroundings; they lived at a time of dissension, wars and general upheaval, at a time when one city (Venice, for example, and the cities under its jurisdiction) would be safe for a progressive scientist while another (such as Rome, or Florence) offered considerable dangers, and when the ideas of a single individual often faced groups of scientists not in agreement with his monomania. To show this, let us look at two astronomers who participated in the development: Copernicus himself and Maestlin, Kepler's teacher.

Copernicus wanted to reform astronomy. He explained his misgivings and the ways in which he tried to overcome them. He wrote:[7]

---

6. The case of Einstein shows that even this modest assumption goes much too far. Einstein recommended a loose opportunism as the best research strategy (see the quotation in the text to footnote 6 of the Introduction) and he warned that a good joke (such as the considerations leading to the special theory of relativity) should not be repeated too often: Philipp Frank, *Einstein, His Life and Times*, London, 1946, p. 261.

7. *Commentariolus*, ed. E. Rosen, *Three Copernican Treatises*, 3rd edition, New York, 1971, translation partly changed in accordance with F. Krafft, 'Copernicus Retroversus I', *Colloquia Copernicana* III and IV, Proceedings of the Joint Symposium of the IAU and the IUHPS, Torun, 1973, p. 119. In what follows I shall also use Krafft, 'Copernicus Retroversus II', loc. cit.

The planetary theories of Ptolemaics and most other astronomers ...
seemed ... to present no small difficulty. For these theories were not
adequate unless certain equants were also conceived; it then appeared that
a planet moved with uniform velocity neither along its own deferent nor
relative to an actual centre. ... Having become aware of these defects I
often considered whether there could perhaps be found a more reasonable
arrangement of circles from which every apparent inequality could be
derived and in which everything would move uniformly about its proper
centre as the rule of accomplished motion requires. ...

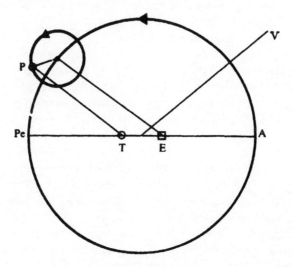

The critique of Copernicus concerns the following model that was used
for calculating the longitudes of Mars, Jupiter and Saturn. The planet P
moves on a small circle, the epicycle, whose centre is located on a larger
circle, the deferent. The centre of the epicycle proceeds with constant
angular velocity with respect to E, the equant point. The planet is
observed from the earth T. E and T are on opposite sides of the centre of
the deferent, having the same distance from it.

Copernicus does not question the empirical adequacy of the model.
On the contrary, he admits that the planetary theories of the Ptolemaeans
and others are 'consistent with the numerical data'.[8] Nor does he believe

---

8. Rosen, op. cit., p. 59.

that these data are in need of correction. Instead of introducing new observations he emphasizes that

> we must follow in their [the ancient Greeks'] footsteps and hold fast to their observations bequeathed to us like an inheritance. And if anyone on the contrary thinks that the ancients are untrustworthy in this regard, surely the gates of this art are closed to him.[9]

Neither new observations nor the inability of Ptolemy to take care of what was known to him are the reason for Copernicus' discomfort. The difficulty he perceives lies elsewhere.

In his account Copernicus distinguishes between absolute motions and apparent motions. The second inequality of planetary motion, i.e. the fact that a planet may run ahead in its path and then reverse its direction, is 'apparent' – it must be reduced to other motions. According to Copernicus these other motions are motions on centred circles with a constant angular velocity around the centre. Ptolemy violates the condition; he uses equants. Equants explain apparent motions not by true motions but again by apparent motions where the planet 'moves with uniform velocity neither along its own deferent, nor relative to an actual centre. . . '. For Copernicus (and for many other astronomers) real motion is a circular motion around a centre with constant angular velocity.[10]

Copernicus removes excentre and equant and replaces them by epicycles.[11] In the Ptolemaic scheme each planet has now three epicycles: the old epicycle and two further epicycles for replacing the excentric and the equant.

In order to avoid this accumulation of epicycles (which occasionally pushed the planets far out into space) Copernicus looks for a different

---

9. *Letter Against Werner*, in Rosen, op. cit., p. 99.

10. Erasmus Reinhold wrote on the title page of his personal copy of *de Revolutionibus: Axioma Astronomicum: Motus coelestis aequalis est et circularis vel ex aequalibus et circularibus compositus.* Quoted from Westman, 'The Melanchthon Circle', op. cit., p. 176.

11. This is true of the *Commentariolus*. In his main work he again uses excentric deferents. Only the equant is replaced by an epicycle. This 'liberation from the equant' (Erasmus Reinhold), also in lunar theory, greatly impressed some of Copernicus' admirers who paid no attention to his new cosmology and the motion of the earth. Westman, 'The Melanchthon Circle, op. cit., pp. 175, 177.

explanation of the second inequality. He is helped by the fact that the second inequality agrees with the position of the sun.[12] It can therefore be interpreted as an apparent motion created by a real (and, of course, circular) motion on part of the earth.

The argument as reconstructed so far (after Krafft) contains two elements: a purely formal element and a reality assertion. Formally it is requested that any periodic motion be reduced to centred circular motions. The request is connected with the assumption that inequalities are apparent while circular motions alone are real. Let us call this *the first reality assumption*. But Copernicus also discovered that his procedure allowed him to incorporate every planetary path into a system, containing the 'large circle', the circle of the earth, as an absolute measure. 'All these phenomena', Copernicus writes in his main work,[13] 'are connected with one another in a most noble way, as if by a golden chain, and each planet with its position and order is a witness that the earth moves while we, who live on the terrestrial globe, failing to recognize its motion, ascribe all sorts of motions to the planets.' It is this inner connectedness of all parts of the planetary system that convinced Copernicus of the reality of the motion of the earth. I call this *the second reality assumption*.

The first reality assumption was part of the Platonic tradition; Aristotle gave it a physical basis. The second reality assumption conflicted with Aristotelian physics and cosmology. Aristotle had already criticized an earlier (Pythagorean) version of it: mathematical harmonies, which are abstractions, reflect truth only if they agree with well-confirmed physical principles. This is a reasonable request; it was used in our own century to reject Schroedinger's interpretation of wave mechanics. It is reasonable especially for those thinkers who regard mathematics as an auxiliary science that may describe but cannot constitute physical processes. It is unreasonable for a Platonist or a Pythagorean. The resulting clash between two interpretations of the nature of mathematical statements played an important role in the 'Copernican Revolution'.

Copernicus strengthened the second reality assumption by referring

---

12. The mean sun in Copernicus. Kepler effects a reduction to the true sun and thereby strengthens the Copernican arrangement.

13. *De Revol.*, Preface to Pope Paul. Krafft assumes that Copernicus discovered this harmony in the course of his attempts to remove the equant and only later turned it into a fundamental argument in favour of a real motion of the earth.

to traditions such as the Hermetic tradition and the idea of the exceptional role of the sun[14] and by showing how it could be reconciled with the phenomena. He made two assumptions. First, that the motion of a body is appropriate to its shape. The earth is spherical, hence its motion must be circular. Secondly, objects such as stones stay with the body (the earth) from which they were separated – hence the falling stone stays close to the tower. According to Aristotle the natural motion of objects, i.e. the upward motion of fire and the downward motion of stones, was determined by the structure of space (central symmetry). According to Copernicus it is determined by the distribution of matter. Copernicus 'saves phenomena' such as the free fall of heavy bodies but provides neither independent arguments nor strict laws that could lead to a detailed comparison. His procedure is *ad hoc*. This does not mean that it is bad; it only means that it cannot be reconciled with the leading methodologies of today.

My second example is Michael Maestlin, Kepler's teacher. Maestlin was an expert astronomer and his judgement was generally respected. He 'only reluctantly abandoned' the Ptolemaic distribution of the spheres – but he was forced to do so by circumstances beyond his control.[15] As far as we can see, the circumstances were, first, the nova of 1572. Maestlin observed it, measured its parallax and put it beyond the sphere of the moon into the sphere of the fixed stars. The first part (beyond the moon) followed for Maestlin from the missing parallax, the second part (fixed stars) from the absence of any proper motion. According to Copernicus, whose ideas Maestlin used at this point, a planet moves more slowly the greater its distance from the sun. Observing the changes of colour and brightness Maestlin (and Tycho, who saw the new star on the way to his alchemical laboratory) inferred that the region above the moon cannot

---

14. The phrase 'and in the middle stands the sun' was not new. In the older astronomy the sun was indeed in the middle of the planets, with Mars, Jupiter and Saturn standing above it, and Venus, Mercury and the moon below. It also 'ruled' the planets in the sense that its motion was mirrored in the motions of all planets (the moon excepted). See e.g. Macrobius, *Somnium Scipionis*.

15. In what follows I am using the dissertation by R.A. Jarrell, *The Life and Scientific Work of the Tuebingen Astronomer Michael Maestlin*, Toronto, 1972, as well as R.S. Westman, 'Michael Maestlin's Adoption of the Copernican Theory', *Colloquia Copernicana* IV, Ossilineum, 1975, p. 53ff.

be without change, as Aristotle had assumed. However, it would be rash to conclude that Maestlin (and Tycho) regarded the nova as a 'blow against the peripatetic philosophy'.[16] Many Church people, Theodore Beza among them, regarded the phenomenon as a return of the star of Bethlehem, i.e. as a supernatural event.[17] Tycho thought this comparison too modest; here, he said, is the greatest miracle since the beginning of the world, comparable at least to Joshua's stopping of the sun.[18] This means that as far as Tycho was concerned miracles refuted the idea of the *autonomy* of the laws of nature (which was an Aristotelian idea), they did not refute *specific laws*. Maestlin, on the other hand, being perhaps more sceptical about miracles, may indeed have regarded the case as a 'blow against' Aristotle.

The next question is how serious a blow it was for him. The idea of a permanent heaven was part of cosmology and contained the special hypothesis of a fifth element. The falsehood of this hypothesis impaired neither the remaining laws of motion nor the tower argument. Both Clavius and Tycho accepted a changing heaven[19] but still used the tower argument to exclude the motion of the earth. If Maestlin's doubts reached further then this was due either to an idiosyncratic interpretation of the Aristotelian doctrines, or to personal inclinations towards a non-Aristotelian world-view. It seems that we must assume the latter.

The next decisive event on Maestlin's journey towards Copernicus was the comet of 1577. Again Maestlin, prompted by 'numerous observations', puts the comet into the superlunar region.[20] The idea that this region is free from change has now definitely been dropped.

Maestlin also tried to determine the trajectory of the comet. He found it to be moving in the path of Venus as described in Book 6, Chapter 12 of *de Revolutionibus*. Somewhat hesitatingly he now accepts the Copernican

---

16. Jarrell, op. cit., p. 108.

17. See the literature in P.H. Kocher, *Science and Religion in Elizabethan England*, New York, 1969, pp. 174f, footnotes 12 and 13. See also Vol. VI, Chapter XXXII of Lynn Thorndike, *A History of Magic and Experimental Science*, New York, 1941.

18. *Progymnasmata*, p. 548.

19. For Clavius, see his commentary on Sacrobosco's sphere, 1593 edition, pp. 210f. See also Westfall, op. cit., p. 44.

20. Jarrell, op. cit., p. 112.

ordering of the spheres.[21] But, so he adds, he was forced to do so 'by extreme necessity'.[22]

This 'extreme necessity' arises only when geometrical considerations are given the force of cosmological arguments. Many years later Galileo cautioned against this way of reasoning: rainbows, he said, cannot be caught by triangulation. Maestlin had no such doubts. He accepted the traditional distinction between physics and astronomy and identified astronomy with mathematics: 'Copernicus wrote his entire book not as a physicist, but as an astronomer' is his comment on the margin of his copy of *de Revolutionibus*.[23] He then interpreted the results of mathematical arguments by using the second reality assumption. This means that he did not *overcome* an Aristotelian resistance against such an interpretation, he acted as if such a resistance *did not exist*. 'This argument', he wrote in his marginal notes,[24] 'is wholly in accord with reason. Such is the arrangement of this entire, immense *machina* that it permits surer demonstrations, indeed, the entire universe revolves in such a way that nothing can be transposed without confusion of its [parts] and, hence, by means of these [surer demonstrations] all the phenomena of motion can be demonstrated most exactly, for nothing unfitting occurs in the course of their orbits.' Kepler too became a Copernican because of this harmony and because of the comet, the interesting fact being that Maestlin's calculations of the path of the comet contain serious mistakes; it did *not* move in the orbit of Venus.

Now let us compare these events and the situations in which they occur with some once popular philosophies of science. We notice at once that none of these philosophies considers all the disciplines that contributed to the debate. Astronomy is in the centre. A rational reconstruction of the developments in this area is thought to be a rational reconstruction of the Copernican Revolution itself. The role of physics (the tower argument), the fact that theology occasionally formed a strong boundary condition (cf. Tycho's reaction to his nova and to the idea of the motion of the earth) and the role of different mathematical philosophies show that this cannot possibly be

21. Ibid., p. 117.
22. Ibid., p. 120.
23. Westman, op. cit., p. 59.
24. Ibid.

true. This fatal incompleteness is the first and most fundamental objection against all reconstructions that have been offered. They still depend on the (positivistic) prejudice that observations alone decide a case and that they can judge a theory all by themselves, without any help (or hindrance) from alternatives, metaphysical alternatives included. Moreover, they even fail in the narrow domain they have chosen for reconstruction, viz. astronomy. To show this, let us consider the following accounts:

1. *Naive empiricism*: the Middle Ages read the Bible and never looked at the sky. Then people suddenly looked upwards and found that the world was different from the opinion of the schools.

This account has disappeared from astronomy – but its analogue survives in other areas (for example, in some parts of the history of medicine). The main argument against it is that Aristotle was an arch empiricist and that Ptolemy used carefully collected data.[25]

2. *Sophisticated empiricism*: new observations forced astronomers to modify an already empirical doctrine.

This certainly is not true for Copernicus and his followers in the 16th century. As we have seen, Copernicus thought the Ptolemaic system to be *empirically adequate* – he criticized it for *theoretical reasons*. And his 'observations' are essentially those of Ptolemy, as he says himself.

Modern comparisons of Copernican and Ptolemaic predictions 'with the facts', i.e. with 19th- and 20th-century calculations, show, furthermore, that empirical predictions were not improved, and actually become worse when the competing systems are restricted to the same number of parameters.[26]

---

25. 'Carefully' has been contested by R.R. Newton, *The Crime of Claudius Ptolemy*, Baltimore, 1977. Newton shows that many of Ptolemy's 'data' *were manufactured* to fit his model. For his optics this has been known for a long time.

26. Stanley E. Babb, 'Accuracy of Planetary Theories, Particularly for Mars', *Isis*, September 1977, pp. 426ff. See also the earlier article of Derek de Solla Price, 'Contra Copernicus', in M. Clagett (ed.), *Critical Problems of the History of Science*, Madison, 1959, pp. 197ff; N.R. Hanson, *Isis*, No. 51, 1960, pp. 150ff, as well as Owen Gingerich, 'Crisis vs Aesthetics in the Copernican Revolution', in Beer (ed.), *Vistas in Astronomy*, Vol. 17, 1974. Gingerich compares the *tables* of Stoeffler, Stadius, Maestlin, Magini and Origanus and finds all of them beset by errors of roughly the same magnitude (though not of the same distribution along the ecliptic).

The only new observations made were those of Tycho Brahe – but they already led beyond Copernicus to Kepler. Galileo's observations belong to cosmology, not to astronomy. They lend plausibility to some of Copernicus' *analogies*. A *compelling* proof of the motion of the earth did not emerge, however, for the Galilean observations could also be accommodated by the Tychonian system.

3. *Falsificationism*: new observations refuted important assumptions of the old astronomy and led to the invention of a new one. This is not correct for Copernicus and the domain of astronomy (see above, comments on 2). The 'refutation' of the immutability of the heavens was neither compelling nor decisive for the problem of the motion of the earth. Besides, the idea of the motion of the earth was in big trouble or, if you will, 'refuted'. It could survive only if it was treated with kindness. But if *it* could be treated with kindness, then so could the older system.

We see here very clearly how misguided it is to try reducing the process 'Copernican Revolution' to a single principle, such as the principle of falsification. Falsifications played a role just as new observations played a role. But both were imbedded in a complex pattern of events which contained tendencies, attitudes, and considerations of an entirely different nature.

4. *Conventionalism*: the old astronomy became more and more complicated – so it was in the end replaced by a simpler theory. It is this assumption that led to the mocking remark of the 'epicyclical degeneration'. The theory overlooks the fact that the Copernican scheme has about as many circles as the Ptolemaic one.[27]

5. *The theory of crises*: astronomy was in a crisis. The crisis led to a revolution which brought about the triumph of the Copernican system.

The answer here is the same as under 2: *empirically* there was no crisis, and no crisis was resolved. A crisis did occur in cosmology, but only *after* the idea of the motion of the earth received a serious hearing. The many complaints about the inexactness of astronomical predictions that

---

27. The reader should consult the very instructive diagrams in de Santillana's edition of Galileo's *Dialogue*, Chicago, 1964.

preceded Copernicus (Regiomontanus, for example) criticized the lack of precise initial conditions and accurate tables, *not* basic theory, and such a criticism would have been quite unjust, as the later examination of these theories shows.[28]

---

28. See footnote 26 above.

The results obtained so far suggest abolishing the distinction between a context of discovery and a context of justification, norms and facts, observational terms and theoretical terms. None of these distinctions plays a role in scientific practice. Attempts to enforce them would have disastrous consequences. Popper's 'critical' rationalism fails for the same reasons.

Let us now use the material of the preceding sections to throw light on the following features of contemporary empiricism: (1) the distinctions between a context of discovery and a context of justification – norms and facts, observational terms and theoretical terms; (2) Popper's 'critical' rationalism; (3) the problem of incommensurability. The last problem will lead us back to the problem of rationality and order vs anarchism, which is the main topic of this essay.

One of the objections which may be raised against my attempt to draw methodological conclusions from historical examples is that it confounds two contexts which are essentially distinct, viz. a context of discovery, and a context of justification. *Discovery* may be irrational and need not follow any recognized method. *Justification*, on the other hand, or – to use the Holy Word of a different school – *criticism*, starts only *after* the discoveries have been made, and it proceeds in an orderly way. 'It is one thing,' writes Herbert Feigl, 'to retrace the historical origins, the psychological genesis and development, the socio-political-economic conditions for the acceptance or rejection of scientific theories; and it is quite another thing to provide a logical reconstruction of the conceptual structure and of the testing of scientific theories.'[1] These are indeed two different *things*, especially as they are done by two different *disciplines* (history of science, philosophy of science), which are quite jealous of their independence. But the question is not what distinctions a fertile mind can dream up when confronted with a complex process, or how some homogeneous material may be subdivided; the question is to what extent the distinction drawn reflects a real difference, and whether science can advance without a strong interaction between the separated

---

1. 'The Orthodox View of Theories', in Radner-Winokur (eds), *Analyses of Theories and Methods of Physics and Psychology*, Minneapolis, 1970, p. 4.

domains. (A river may be subdivided by national boundaries but this does not make it a discontinuous entity.) Now there is, of course, a very noticeable difference between the rules of testing as 'reconstructed' by philosophers of science and the procedures which scientists use in actual research. This difference is apparent to the most superficial examination. On the other hand, a most superficial examination also shows that a determined application of the methods of criticism and proof which are said to belong to the context of justification would wipe out science as we know it – and would never have permitted it to arise.[2] Conversely, the fact that science exists proves that these methods were frequently overruled. They were overruled by procedures which belong to the context of discovery. Thus the attempt 'to retrace the historical origins, the psychological genesis and development, the socio-political-economic conditions for the acceptance or rejection of scientific theories', far from being irrelevant for the standards of test, actually leads to a criticism of these standards – *provided* the two domains, historical research and discussion of test procedures, are not kept apart by fiat.

In another paper Feigl repeats his arguments and adds some further points. He is 'astonished that . . . scholars such as N.R. Hanson, Thomas Kuhn, Michael Polanyi, Paul Feyerabend, Sigmund Koch *et al.*, consider the distinction as invalid or at least misleading'.[3] And he points out that neither the psychology of invention nor any similarity, however great, between the sciences and the arts can show that it does not exist. In this he is certainly right. Even the most surprising stories about the manner in which scientists arrive at their theories cannot exclude the possibility that they proceed in an entirely different way once they have found them. *But this possibility is never realized.* Inventing theories and contemplating them in a relaxed and 'artistic' fashion, scientists often make moves that are forbidden by methodological rules. For example, they interpret the evidence so that it fits their fanciful ideas, eliminate difficulties by *ad hoc* procedures, push them aside, or simply refuse to take them seriously. The activities which according to Feigl belong to the context of discovery are, therefore, not just *different* from what philosophers say about justification, *they are in conflict with it.* Scientific practice does not contain two contexts moving *side by side*, it is a complicated *mixture* of

---

2. See the examples in Chapter 5.
3. 'Empiricism at Bay', MS, 1972, p. 2.

procedures, and we are faced by the question if this mixture should be left as it is, or if it should be replaced by a more 'orderly' arrangement. This is part one of the argument. Now we have seen that science as we know it today could not exist without a frequent overruling of the context of justification. This is part two of the argument. The conclusion is clear. Part one shows that we do not have a difference, but a mixture. Part two shows that replacing the mixture by an order that contains discovery on one side and justification on the other would have ruined science: we are dealing with a uniform practice all of whose ingredients are equally important for the growth of science. This disposes of the distinction.

A similar argument applies to the ritual distinction between methodological *prescriptions* and historical *descriptions*. Methodology, it is said, deals with what *should* be done and cannot be criticized by reference to *what is*. But we must of course make sure that our prescriptions have a *point of attack* in the historical material, and we must also make sure that their determined application leads to desirable results. We make sure by considering (historical, sociological, physical, psychological, etc.) *tendencies and laws* which tell us what is possible and what is not possible under the given circumstances and thus separate feasible prescriptions from those which are going to lead into dead ends. Again, progress can be made only if the distinction between the *ought* and the *is* is regarded as a temporary device rather than as a fundamental boundary line.

A distinction which once may have had a point but which has now definitely lost it is the distinction between *observational* terms and *theoretical* terms. It is now generally admitted that this distinction is not as sharp as it was thought to be only a few decades ago. It is also admitted, in complete agreement with Neurath's original views, that *both* theories *and* observations can be abandoned: theories may be removed because of conflicting observations, observations may be removed for theoretical reasons. Finally, we have discovered that *learning* does not go from observation to theory but always involves both elements. Experience arises *together* with theoretical assumptions not before them, and an experience without theory is just as incomprehensible as is (allegedly) a theory without experience: eliminate part of the theoretical knowledge of a sensing subject and you have a person who is completely disoriented and incapable of carrying out the simplest action. Eliminate further knowledge and his sensory world (his 'observation language') will start disintegrating, colours and other simple sensations will disappear, until he is in a stage even more primitive than a small child. A small child, on the other hand, does not possess a stable perceptual world which he uses for making sense of the theories put before him. Quite the contrary – he passes through various perceptual stages which are only loosely connected with each other (earlier stages *disappear* when new stages take over – see Chapter 16) and which embody all the theoretical knowledge available at the time. Moreover, the whole process starts only because the child reacts correctly towards signals, *interprets them correctly*, because he possesses means of interpretation even before he has experienced his first clear sensation.

All these discoveries cry out for a new terminology that no longer separates what is so intimately connected in the development both of the individual and of science at large. Yet the distinction between observation and theory is still upheld. But what is its point? Nobody will deny that the sentences of science can be classified into long sentences and short sentences, or that its statements can be classified into those which are intuitively obvious and others which are not. Nobody will deny that such distinctions *can be made*. But nobody will put great weight on them, or will even mention them, *for they do not now play any decisive role in the business of science*. (This was not always so. Intuitive plausibility, for example, was once thought to be a most important guide to the truth; it disappeared from methodology the very moment intuition was replaced by experience, and by formal considerations.) Does experience play such a role? It does not, as we have seen. Yet the inference that the distinction between theory and observation has now ceased to be relevant, is either not drawn or is explicitly rejected.[4] Let us take a step forward and let us abandon this last trace of dogmatism in science!

Incommensurability, which I shall discuss next, is closely connected with the question of the rationality of science. Indeed one of the most general objections not merely to the *use of* incommensurable theories but even to the idea that *there are* such theories to be found in the history of science is the fear that they would severely restrict the efficacy of traditional, non-dialectical *argument*. Let us, therefore, look a little more closely at the critical *standards* which, according to some, constitute

---

4. 'Neurath fails to give . . . rules [which distinguish empirical statements from others] and thus unwittingly throws empiricism overboard', K.R. Popper, *The Logic of Scientific Discovery*, New York and London, 1959, p. 97.

the content of a 'rational' argument. More especially, let us look at the standards of the Popperian school, which are still being taken seriously in the more backward regions of knowledge. This will prepare us for the final step in our discussion of the issue between law-and-order methodologies and anarchism in science.

Some readers of my arguments in the above text have pointed out that Popper's 'critical' rationalism is sufficiently liberal to accommodate the developments I have described. Now critical rationalism is either a meaningful idea or it is a collection of slogans that can be adapted to any situation.

In the first case it must be possible to produce rules, standards, restrictions which permit us to separate critical behaviour (thinking, singing, writing of plays) from other types of behaviour so that we can *discover* irrational actions and *correct* them with the help of concrete suggestions. It is not difficult to produce the standards of rationality defended by the Popperian school.

These standards are standards of *criticism*: rational discussion consists in the attempt to criticize, and not in the attempt to prove or to make probable. Every step that protects a view from criticism, that makes it safe or 'well-founded', is a step away from rationality. Every step that makes it more vulnerable is welcome. In addition, it is recommended to abandon ideas which have been found wanting and it is forbidden to retain them in the face of strong and successful criticism unless one can present suitable counter-arguments. Develop your ideas so that they can be criticized; attack them relentlessly; do not try to protect them, but exhibit their weak spots; eliminate them as soon as such weak spots have become manifest – these are some of the rules put forth by our critical rationalists.

These rules become more definite and more detailed when we turn to the philosophy of science and, especially, to the philosophy of the natural sciences.

Within the natural sciences, criticism is connected with experiment and observations. The content of a theory consists in the sum total of those basic statements which contradict it; it is the class of its potential falsifiers. Increased content means increased vulnerability, hence theories of large content are to be preferred to theories of small content. Increase of content is welcome, decrease of content is to be avoided. A theory that contradicts an accepted basic statement must be given up. *Ad hoc* hypotheses are forbidden – and so on. A science, however, that accepts the rules of a critical empiricism of this kind will develop in the following manner.

We start with a *problem*, such as the problem of the planets at the time of Plato. This problem (which I shall discuss in a somewhat idealized form) is not merely the result of *curiosity*, it is a *theoretical result*. It is due to the fact that certain *expectations* have been disappointed: on the one hand it seems to be clear that the stars must be divine, hence one expects them to behave in an orderly and lawful manner. On the other hand, one cannot find any easily discernible regularity. The planets, to all intents and purposes, move in a quite chaotic fashion. How can this fact be reconciled with the expectation and with the principles that underlie the expectation? Does it show that the expectation is mistaken? Or have we failed in our analysis of the facts? This is the problem.

It is important to see that the elements of the problem are not simply *given*. The 'fact' of irregularity, for example, is not accessible without further ado. It cannot be discovered by just anyone who has healthy eyes and a good mind. It is only through a certain expectation that it becomes an object of our attention. Or, to be more accurate, this fact of irregularity *exists* because there is an expectation of regularity and because there are ideas which define what it means to be 'regular'. After all, the term 'irregularity' makes sense only if we have a rule. In our case the rule that defines regularity asserts circular motion with constant angular velocity. The fixed stars agree with this rule and so does the sun, if we trace its path relative to the fixed stars. The planets do not obey the rule, neither directly, with respect to the earth, nor indirectly, with respect to the fixed stars.

(In the problem we are examining now the rule is formulated explicitly and it can be discussed. This is not always the case. Recognizing a colour as red is made possible by deep-lying patterns concerning the structure of our surroundings, and recognition does not occur when these patterns cease to exist.)

To sum up this part of the Popperian doctrine: research starts with a problem. The problem is the result of a conflict between an expectation and an observation which is constituted by the expectation. It is clear that this doctrine differs from the doctrine of inductivism where objective facts enter a passive mind and leave their traces there. It was prepared by Kant, Mach, Poincaré, Dingler, and by Mill (*On Liberty*).

Having formulated a problem, one tries to *solve* it. Solving a problem means inventing a theory that is relevant, falsifiable (to a degree larger than any alternative), but not yet falsified. In the case mentioned above (planets at the time of Plato), the problem is: to find circular motions of constant angular velocity for the purpose of saving the planetary phenomena. A first solution was provided by Eudoxos and then by Heracleides of Pontos.

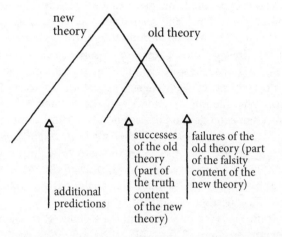

Next comes the *criticism* of the theory that has been put forth in the attempt to solve the problem. Successful criticism removes the theory *once and for all* and creates a new problem, viz. to explain (a) why the theory was successful so far; (b) why it failed. Trying to solve *this* problem we need a new theory that reproduces the successful consequences of the older theory, denies its mistakes and makes additional predictions not made before. These are some of the *formal conditions* which a *suitable successor of a refuted theory* must satisfy. Adopting the conditions, one proceeds by conjecture and refutation from less general theories to more general theories and expands the content of human knowledge.

More and more facts are *discovered* (or constructed with the help of expectations) and are then explained by theories. There is no guarantee that scientists will solve every problem and replace every theory that has been refuted with a successor satisfying the formal conditions. The invention of theories depends on our talents and other fortuitous circumstances such as a satisfactory sex life. But as long as these talents hold out, the enclosed scheme is a correct account of the growth of a knowledge that satisfies the rules of critical rationalism.

Now at this point, one may raise two questions.
1. Is it *desirable* to live in accordance with the rules of a critical rationalism?
2. Is it *possible* to have both a science as we know it and these rules?

As far as I am concerned, the first question is far more important than the second. True, science and related institutions play an important part in our culture, and they occupy the centre of interest for many philosophers (most philosophers are opportunists). Thus the ideas of the Popperian school were obtained by generalizing solutions for methodological and epistemological problems. Critical rationalism arose from the attempt to understand the Einsteinian revolution, and it was then extended to politics and even to the conduct of one's private life. Such a procedure may satisfy a *school philosopher*, who looks at life through the spectacles of his own technical problems and recognizes hatred, love, happiness, only to the extent to which they occur in these problems. But if we consider human interests and, above all, the question of human freedom (freedom from hunger, despair, from the tyranny of constipated systems of thought and *not* the academic 'freedom of the will'), then we are proceeding in the worst possible fashion.

For is it not possible that science as we know it today, or a 'search for the truth' in the style of traditional philosophy, will create a monster? Is it not possible that an objective approach that frowns upon personal connections between the entities examined will harm people, turn them into miserable, unfriendly, self-righteous mechanisms without charm and humour? 'Is it not possible,' asks Kierkegaard, 'that my activity as an objective [or a critico-rational] observer of nature will weaken my strength as a human being?'[5] I suspect the answer to many of these questions is affirmative and I believe that a reform of the sciences that makes them more anarchic and more subjective (in Kierkegaard's sense) is urgently needed.

---

5. *Papirer*, ed. Heiberg, VII, Pt. I, sec. A, No. 182. Mill tries to show how scientific method can be understood as part of a theory of man, and thus gives a positive answer to the question raised by Kierkegaard; see footnote 2 to Chapter 4.

But these are not the problems I want to discuss now. In the present essay I shall restrict myself to the second question and I shall ask: Is it possible to have both a science as we know it and the rules of a critical rationalism as just described? And to *this* question the answer seems to be a firm and resounding NO.

To start with we have seen, though rather briefly, that the actual development of institutions, ideas, practices, and so on, often *does not start from a problem* but rather from some extraneous activity, such as playing, which, as a side effect, leads to developments which later on can be interpreted as solutions to unrealized problems.[6] Are such developments to be excluded? And, if we do exclude them, will this not considerably reduce the number of our adaptive reactions and the quality of our learning process?

Secondly, we have seen, in Chapters 8–14, that a *strict principle of falsification*, or a 'naive falsificationism' as Lakatos calls it,[7] would wipe out science as we know it and would never have permitted it to start.

The demand for *increased content* is not satisfied either. Theories which effect the overthrow of a comprehensive and well-entrenched point of view, and take over after its demise, are initially restricted to a fairly narrow domain of facts, to a series of paradigmatic phenomena which lend them support, and they are only slowly extended to other areas. This can be seen from historical examples (footnote 12 of Chapter 8), and it is also plausible on general grounds: trying to develop a new theory, we must first take a *step back* from the evidence and reconsider the problem of observation (this was discussed in Chapter 11). Later on, of course, the theory is extended to other domains; but the mode of extension is only rarely determined by the elements that constitute the content of its predecessors. The slowly emerging conceptual apparatus of the theory *soon starts defining its own problems*, and earlier problems, facts, and observations are either forgotten or pushed aside as irrelevant. This is an entirely natural development, and quite unobjectionable. For

---

6. See the brief comments on the relation between idea and action in Chapter 1. For details see footnotes 31ff of 'Against Method', *Minnesota Studies*, Vol. 4, 1970.

7. 'Falsification and the Methodology of Scientific Research Programmes', in Lakatos and Musgrave (eds), *Criticism and the Growth of Knowledge*, Cambridge, 1970, pp. 93ff. ('Naive falsificationism' is here also called 'dogmatic'.)

why should an ideology be constrained by older problems which, at any rate, make sense only in the abandoned context and which look silly and unnatural now? Why should it even *consider* the 'facts' that gave rise to problems of this kind or played a role in their solutions? Why should it not rather proceed in its own way, devising its own tasks and assembling its own domain of 'facts'? A comprehensive theory, after all, is supposed to contain also an *ontology* that determines what exists and thus delimits the domain of possible facts and possible questions. The development of science agrees with these considerations. New views soon strike out in new directions and frown upon the older *problems* (What is the base upon which the earth rests? What is the specific weight of phlogiston? What is the absolute velocity of the earth?) and the older *facts* (most of the facts described in the *Malleus Maleficarum* – Chapter 8, footnote 2 – the facts of Voodoo – Chapter 4, footnote 8 – the properties of phlogiston or those of the ether) which so much exercised the minds of earlier thinkers. And where they *do* pay attention to preceding theories, they try to accommodate their factual core in the manner already described, with the help of *ad hoc* hypotheses, *ad hoc* approximations, redefinition of terms, or by simply *asserting*, without any more detailed study of the matter, that the core 'follows from' the new basic principles.[8] They are 'grafted on to older programmes with which they [are] blatantly inconsistent.'[9]

The result of all these procedures is an interesting *epistemological illusion*: the *imagined* content of the earlier theories (which is the intersection of the remembered consequences of these theories with the newly recognized domain of problems and facts) *shrinks* and may decrease to such an extent that it becomes smaller than the *imagined* content of the new ideologies (which are the actual consequences of these ideologies *plus* all those 'facts', laws, principles which are tied to them by *ad hoc* hypotheses, *ad hoc* approximations or by the say-so of some influential physicist or philosopher of science – and which properly belong to the predecessor). Comparing the old and the new it thus *appears* that the relation of empirical contents is like this

---

8. 'Einstein's theory is better than . . . Newton's theory *anno 1916* . . . *because* it explained everything that Newton's theory had successfully explained . . .', Lakatos, op. cit., p. 214.

9. Lakatos, discussing Copernicus and Bohr, ibid., p. 143.

new

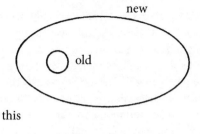

or, perhaps, like this

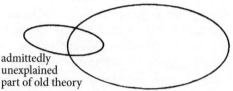

admittedly
unexplained
part of old theory

while in actual fact it is much more like this

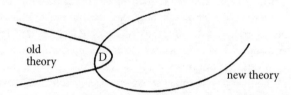

old
theory   D

new theory

domain D representing the problems and facts of the old theory which are still remembered and which have been distorted so as to fit into the new framework. It is this illusion which is responsible for the persistent survival of the demand for increased content.[10]

_____

10. This illusion is the core of Elie Zahar's excellent paper on the development from Lorentz to Einstein. According to Zahar, Einstein superseded Lorentz with the explanation of the perihelion of Mercury (1915). But in 1915 nobody had as yet succeeded in giving a relativistic account of classical perturbation theory to the degree of approximation reached by Laplace and Poincaré, and the implications of Lorentz on the atomic level (electron theory of metals) were not accounted for either, but were gradually replaced by the quantum theory: Lorentz was 'superseded' not by one, but by at least two different and mutually incommensurable programmes. Lakatos, in his excellent reconstruction of the development of the research programme of Copernicus from the *Commentariolus* to the *De Revol.*, notes progressive changes but only because he omits the dynamical and the optical problems and concentrates on kinematics, pure and simple. Small wonder that both Zahar and Lakatos are under the impression that the content condition is still satisfied. See my short note 'Zahar on Einstein', in the *British Journal for the Philosophy of Science*, March

Finally, we have by now seen quite distinctly the need for *ad hoc hypotheses*: *ad hoc* hypotheses and *ad hoc* approximations create a tentative area of contact between 'facts' and those parts of a new view which seem capable of explaining them, at some time in the future and after addition of much further material. They specify possible explananda and explanatia, and thus determine the direction of future research. They may have to be retained forever if the new framework is partly unfinished (this happened in the case of the quantum theory, which needs the classical concepts to turn it into a complete theory). Or they are incorporated into the new theory as theorems, leading to a redefinition of the basic terms of the proceeding ideology (this happened in the cases of Galileo and of the theory of relativity). The demand that the truth-content of the earlier theory as *conceived while the earlier theory reigned supreme* be included in the truth-content of the successor is violated in either case.

To sum up: wherever we look, whatever examples we consider, we see that the principles of critical rationalism (take falsifications seriously; increase content; avoid *ad hoc* hypotheses; 'be honest' – whatever *that* means; and so on) and, *a fortiori*, the principles of logical empiricism (be precise; base your theories on measurements; avoid vague and untestable ideas; and so on), though practised in special areas, give an inadequate account of the past development of science as a whole and are liable to hinder it in the future. They give an inadequate account of science because science is much more 'sloppy' and 'irrational' than its methodological image. And they are liable to hinder it because the attempt to make science more 'rational' and more precise is bound to wipe it out, as we have seen. The difference between science and methodology which is such an obvious fact of history, therefore, indicates a weakness of the latter, and perhaps of the 'laws of reason' as well. For what appears as 'sloppiness', 'chaos' or 'opportunism' when compared with such laws has a most important function in the development of those very theories which we today regard as essential parts of our knowledge of nature. *These 'deviations', these 'errors', are preconditions of progress.* They permit knowledge to survive in the complex and difficult world which we inhabit, they permit *us* to remain free and happy agents. Without 'chaos', no knowledge. Without a frequent dismissal of reason, no progress. Ideas

---

1974 as well as R.N. Nugaev, 'Special Relativity as a Stage in the Development of Quantum Theory', *Historia Scientiarum*, No. 34, 1988, pp. 57ff.

which today form the very basis of science exist only because there were such things as prejudice, conceit, passion; because these things *opposed reason*; and because they *were permitted to have their way*. We have to conclude, then, that *even within* science reason cannot and should not be allowed to be comprehensive and that it must often be overruled, or eliminated, in favour of other agencies. There is not a single rule that remains valid under all circumstances and not a single agency to which appeal can always be made.[11]

11. Even Lakatos' ingenious methodology does not escape this indictment. Lakatos seems liberal because he forbids very little and he seems rational because he still forbids something. But the only thing he forbids is to *describe* a 'degenerating research programme', i.e. a research programme lacking in novel predictions and cluttered with *ad hoc* adaptations, as progressive. He does not forbid its use. But this means that his standards permit a criminal to commit as many crimes as he wants provided he never lies about them. Details in my *Philosophical Papers*, Vol. 2, Chapter 10.

Appendix 1

Having listened to one of my anarchistic sermons, Professor Wigner exclaimed: 'But surely, you do not read all the manuscripts which people send you, but you throw most of them into the wastepaper basket.' I most certainly do. 'Anything goes' does not mean that I shall read every single paper that has been written – God forbid! – it means that I make my selection in a highly individual and idiosyncratic way, partly because I can't be bothered to read what doesn't interest me – and my interests change from week to week and day to day – partly because I am convinced that humanity and even Science will profit from everyone doing his own thing: a physicist might prefer a sloppy and partly incomprehensible paper full of mistakes to a crystal-clear exposition because it is a natural extension of his own, still rather disorganized, research and he might achieve success as well as clarity long before his rival who has vowed never to read a single woolly line (one of the assets of the Copenhagen School was its ability to avoid premature precision). On other occasions he might look for the most perfect proof of a principle he is about to use in order not to be sidetracked in the debate of what he considers to be his main results. There are of course so-called 'thinkers' who subdivide their mail in exactly the same way, come rain, come sunshine, and who also imitate each other's principles of choice – but we shall hardly admire them for their uniformity, and we shall certainly not think their behaviour 'rational'. Science needs people who are adaptable and inventive, not rigid imitators of 'established' behavioural patterns.

In the case of institutions and organizations such as the National Science Foundation the situation is exactly the same. The physiognomy of an organization and its efficiency depends on its members and it improves with their mental and emotional agility. Even Procter and Gamble realized that a bunch of yes-men is inferior in competitive potential to a group of people with unusual opinions, and business has found ways of incorporating the most amazing nonconformists into their machinery. Special problems arise with foundations that distribute money and want to do this in a just and reasonable way. Justice seems to demand that the allocation of funds be carried out on the basis of standards which do not change from one applicant to the next and which reflect the intellectual situation in the fields to be supported. The demand can be satisfied in an *ad hoc* manner without appeal to *universal* 'standards of rationality':

any free association of people must respect the illusions of its members and must give them institutional support. The illusion of *rationality* becomes especially strong when a scientific institution opposes political demands. In this case one class of standards is set against another such class – and this is quite legitimate: each organization, each party, each religious group has a right to defend its particular form of life and all the standards it contains. *But scientists go much further.* Like the defenders of The One True Religion before them they insinuate that their standards are essential for arriving at the Truth, or for getting Results, and they deny such authority to the demands of the politician. They oppose all political interference, and they fall over each other trying to remind the listener, or the reader, of the disastrous outcome of the Lysenko affair.

Now we have seen that the belief in a unique set of standards that has always led to success and will always lead to success is nothing but a chimaera. The *theoretical* authority of science is much smaller than it is supposed to be. Its *social* authority, on the other hand, has by now become so overpowering *that political interference is necessary to restore a balanced development*. And to judge the *effects* of such interference one must study more than one unanalysed case. One must remember those cases where science, left to itself, committed grievous blunders, and one must not forget the instances when political interference did *improve* the situation.[1] Such a balanced presentation of the evidence may even convince us that the time is overdue for adding the separation of state and science to the by now quite customary separation of state and church. Science is only *one* of the many instruments people invented to cope with their surroundings. It is not the only one, it is not infallible, and it has become too powerful, too pushy, and too dangerous to be left on its own. Next, a word about the *practical aim* rationalists want to realize with the help of their methodology.

Rationalists are concerned about intellectual pollution. I share this concern. Illiterate and incompetent books flood the market, empty verbiage full of strange and esoteric terms claims to express profound insights, 'experts' without brains, character, and without even a modicum of intellectual, stylistic, emotional temperament tell us about our 'condition' and the means for improving it, and they do not only preach to us who might be able to see through them, they are let loose on our

---

1. An example was discussed in the text to footnotes 9–12 of Chapter 4.

children and permitted to drag them down into their own intellectual squalor.[2] 'Teachers' using grades and the fear of failure mould the brains of the young until they have lost every ounce of imagination they might once have possessed. This is a disastrous situation, and one not easily mended. But I do not see how a rationalistic methodology can help. As far as I am concerned the first and the most pressing problem is to get education out of the hands of the 'professional educators'. The constraints of grades, competition, regular examination must be removed and *we must also separate the process of learning from the preparation for a particular trade.* I grant that business, religions, special professions such as science or prostitution, have a right to demand that their participants and/or practitioners conform to standards they regard as important, and that they should be able to ascertain their competence. I also admit that this implies the need for special types of education that prepare a man or a woman for the corresponding 'examinations'. The standards taught need not be 'rational' or 'reasonable' in any sense, though they will be usually presented as such; it suffices that they are *accepted* by the groups one wants to join, be it now Science, or Big Business, or The One True Religion. After all, in a democracy 'reason' has just as much right to be heard and to be expressed as 'unreason', especially in view of the fact that one man's 'reason' is the other man's insanity. But one thing must be avoided at all costs: the special standards which define special subjects and special professions must not be allowed to permeate *general* education and they must not be made the defining property of a 'well-educated person'. General education should prepare citizens to *choose between* the standards, or to find their way in a society that contains groups committed to various standards, *but it must under no condition bend their minds so that they conform to the standards of one particular group.* The standards will be *considered,* they will be *discussed,* children will be encouraged to get proficiency in the more important subjects, *but only as one gets proficiency in a game,* that is, without serious commitment and without robbing the mind of its ability to play other games as well. Having been prepared in this way a young person may decide to devote the rest of his life to a particular profession and he may start taking it seriously forthwith. This 'commitment' should be the result

---

2. Even the law now seems to support these tendences, as is shown in Peter Huber's *Galileo's Revenge,* New York, 1991.

of a conscious decision, on the basis of a fairly complete knowledge of alternatives, *and not a foregone conclusion.*

All this means, of course, that we must stop the scientists from taking over education and from teaching as 'fact' and as 'the one true method' whatever the myth of the day happens to be. Agreement with science, decision to work in accordance with the canons of science should be the result of examination and choice, and *not* of a particular way of bringing up children.

It seems to me that such a change in education and, as a result, in perspective will remove a great deal of the intellectual pollution rationalists deplore. The change of perspective makes it clear that there are many ways of ordering the world that surrounds us, that the hated constraints of one set of standards may be broken by freely accepting standards of a different kind, and that there is no need to reject *all* order and to allow oneself to be reduced to a whining stream of consciousness. A society that is based on a set of well-defined and restrictive rules, so that being human becomes synonymous with obeying these rules, *forces the dissenter into a no-man's-land of no rules at all and thus robs him of his reason and his humanity.* It is the paradox of modern irrationalism that its proponents silently identify rationalism with order and articulate speech and thus see themselves forced to promote stammering and absurdity – many forms of 'mysticism' and 'existentialism' are impossible without a firm but unrealized commitment to some principles of the despised ideology (just remember the 'theory' that poetry is nothing but emotions colourfully expressed). Remove the principles, admit the possibility of many different forms of life, and such phenomena will disappear like a bad dream.

My diagnosis and my suggestions coincide with those of Lakatos – up to a point. Lakatos has identified overly-rigid rationality principles as the source of some versions of irrationalism and he has urged us to adopt new and more liberal standards. I have identified overly-rigid rationality principles as well as a general respect for 'reason' as the source of some forms of mysticism and irrationalism, and I also urge the adoption of more liberal standards. But while Lakatos' great 'respect for great science'[3] makes him look for the standards within the confines of modern science 'of the last two centuries,'[4] I recommend to put science in its place

---

3. 'History', p. 113.
4. Ibid., p. 111.

as an interesting but by no means exclusive form of knowledge that has many advantages but also many drawbacks: 'Although science taken as a whole is a nuisance, one can still learn from it.'[5] Also I don't believe that charlatans can be banned just by tightening up rules.

Charlatans have existed at all times and in the most tightly-knit professions. Some of the examples which Lakatos mentions[6] seem to indicate that the problem is created by too much control and not by too little.[7] This is especially true of the new 'revolutionaries' and their 'reform' of the universities. Their fault is that they are puritans and *not* that they are libertines.[8] Besides, who would expect that cowards will improve the intellectual climate more readily than will libertines? (Einstein saw this problem and he therefore advised people not to connect their research with their profession: research has to be free from the pressures which professions are likely to impose.[9]) We must also remember that those rare cases where liberal methodologies *do* encourage empty verbiage and loose thinking ('loose' from one point of view, though perhaps not from another) may be inevitable in the sense that the guilty liberalism is *also* a precondition of a free and humane life.

Finally, let me repeat that for me the chauvinism of science is a much greater problem than the problem of intellectual pollution. It may even be one of its major causes. Scientists are not content with running their own playpens in accordance with what they regard as the rules of scientific method, they want to universalize these rules, they want them to become part of society at large and they use every means at their disposal – argument, propaganda, pressure tactics, intimidation, lobbying – to achieve their aims. The Chinese Communists recognized the dangers inherent in this chauvinism and they proceeded to remove it. In the process they restored important parts of the intellectual and emotional heritage of the Chinese people and they also improved the practice of medicine.[10] It would be of advantage if other governments followed suit.

---

5. Gottfried Benn, letter to Gert Micha Simon of 11 October 1949, quoted from Gottfried Benn, *Lyrik und Prosa, Briefe und Dokumente*, Wiesbaden, 1962, p. 235.

6. 'Falsification', p. 176, footnote 1.

7. See also his remarks on 'false consciousness' in 'History', pp. 94, 108ff.

8. For an older example, cf. the *Born–Einstein Letters*, New York, 1971, p. 150.

9. Ibid., pp. 105ff.

10. See text to footnotes 9–12 of Chapter 4.

16

Finally, the kind of comparison that underlies most methodologies is possible only in some rather simple cases. It breaks down when we try to compare non-scientific views with science or when we consider the most advanced, most general and therefore most mythological parts of science itself.

I have much sympathy with the view, formulated clearly and elegantly by Whorf (and anticipated by Bacon), that languages and the reaction patterns they involve are not merely instruments for *describing* events (facts, states of affairs), but that they are also *shapers* of events (facts, states of affairs),[1] that their 'grammar' contains a cosmology, a comprehensive view of the world, of society, of the situation of man[2] which influences thought, behaviour, perception.[3] According to Whorf the cosmology of a language is expressed partly by the overt use of words, but it also rests on classifications 'which ha[ve] no overt mark . . . but which operate [. . .] through an invisible "central exchange" of linkage bonds in such a way as to determine other words which mark the class.'[4] Thus '[t]he gender nouns, such as boy, girl, father, wife, uncle, woman, lady, including thousands of given names like George, Fred, Mary, Charlie, Isabel, Isadore, Jane, John, Alice, Aloysius, Esther, Lester, bear no distinguishing mark of gender like the Latin *-us* or *-a* within each motor process, but nevertheless each of these thousands of words has an invariable linkage bond connecting it with absolute precision either to the word "he" or to the word "she" which, however, does not come into the overt behaviour picture until and unless special situations of discourse require it.'[5]

---

1. According to Whorf 'the background linguistic system (in other words, the grammar) of each language is not merely a reproducing system for voicing ideas, but rather is itself a shaper of ideas, the programme and guide for the individual's mental activity, for his analysis of impressions, for his synthesis of his mental stock in trade'. *Language, Thought and Reality*, Cambridge, Mass., 1956, p. 121. See also Appendix 2.

2. As an example see Whorf's analysis of Hopi Metaphysics in ibid., pp. 57ff.

3. 'Users of markedly different grammars are pointed by their grammars towards different types of observations . . .', Ibid., p. 221.

4. Ibid., p. 69.

5. Ibid., p. 68.

Covert classifications (which, because of their subterranean nature, are 'sensed rather than comprehended – awareness of [them] has an intuitive quality'[6] – which 'are quite apt to be more rational than overt ones'[7] and which may be very 'subtle' and not connected 'with any grand dichotomy'[8]) create 'patterned resistances to widely divergent points of view'.[9] If these resistances oppose not just the truth of the resisted alternatives but the presumption that an alternative has been presented, then we have an instance of incommensurability.

I also believe that scientific theories, such as Aristotle's theory of motion, the theory of relativity, the quantum theory, classical and modern cosmology are sufficiently general, sufficiently 'deep' and have developed in sufficiently complex ways to be considered along the same lines as natural languages. The discussions that prepare the transition to a new age in physics, or in astronomy, are hardly ever restricted to the overt features of the orthodox point of view. They often reveal hidden ideas, replace them by ideas of a different kind, and change overt as well as covert classifications. Galileo's analysis of the tower argument led to a clearer formulation of the Aristotelian theory of space and it also revealed

---

6. Ibid., p. 70. Even '[a] phoneme may assume definite semantic duties as part of its rapport. In English the phoneme ð ['thorn'] (the voiced sound of *th*) occurs initially only in the cryptotype [covert classification not connected with any grand dichotomy – p. 70] of demonstrative particles (the, this, there, than, etc.). Hence, there is a *psychic pressure* against accepting the voiced sound of *th* in new or imaginary words: *th*ig, *th*ay, *th*ob, *th*uzzle, etc., not having demonstrative meaning. Encountering such a new word (e.g. *th*ob) on a page, we will "instinctively" give it the voiceless sound θ of *th* in "think". But it is not "instinct". Just our old friend linguistic rapport again' (p. 76, my italics).

7. Ibid., p. 80. The passage continues: '. . . some rather formal and not very meaningful linguistic group, marked by some overt feature, may happen to coincide very roughly with some concatenation of phenomena in such a way as to suggest a rationalization of this parallelism. In the course of phonetic change, the distinguishing mark, ending, or what not is lost, and the class passes from a formal to a semantic one. Its reactance is now what distinguishes it as a class, and its idea is what unifies it. As time and use go on, it becomes increasingly organized around a rationale, it attracts semantically suitable words and loses former members that now are semantically inappropriate. Logic is now what holds it together.' See also Mill's account of his educational development as described in text to footnote 14 of Chapter 11.

8. Whorf, op. cit., p. 70. Such subtle classifications are called cryptotypes by Whorf. A cryptotype is 'a submerged, subtle, and elusive meaning, corresponding to no actual word, yet shown by linguistic analysis to be functionally important in the grammar'.

9. Ibid., p. 247.

the difference between impetus (an absolute magnitude that inheres in the object) and momentum (which depends on the chosen reference system). Einstein's analysis of simultaneity unearthed some features of the Newtonian cosmology which, though unknown, had influenced all arguments about space and time, while Niels Bohr found in addition that the physical world could not be regarded as being entirely separated from the observer and thus gave content to the idea of independence that was part of classical physics. Attending to cases such as these we realize that scientific arguments may indeed be subjected to 'patterned resistances' and we expect that incommensurability will also occur among theories.

(As incommensurability depends on covert classifications and involves major conceptual changes it is hardly ever possible to give an explicit definition of it. Nor will the customary 'reconstructions' succeed in bringing it to the fore. The phenomenon must be shown, the reader must be led up to it by being confronted with a great variety of instances, and he must then judge for himself. This will be the method adopted in the present chapter.)

Interesting cases of incommensurability occur already in the domain of *perception*. Given appropriate stimuli, but different systems of classification (different 'mental sets'), our perceptual apparatus may produce perceptual objects which cannot be easily compared.[10] A direct judgement is impossible. We may compare the two objects in our *memory*, but *not* while attending to the *same picture*. The first drawing, above, goes

---

10. 'A master of introspection, Kenneth Clark, has recently described to us most vividly how even he was defeated when he attempted to "stall" an illusion. Looking at a great Velásquez, he wanted to observe what went on when the brush strokes and dabs of pigment on the canvas transformed themselves into a vision of transfigured reality as he stepped back. But try as he might, stepping backward and forward, he could never hold both visions at the same time . . ', E. Gombrich, *Art and Illusion*, Princeton, 1956, p. 6.

one step further. It gives rise to perceptual objects which do not just *negate* other perceptual objects – thus retaining the basic categories – but prevent the formation of any object whatsoever (note that the cylinder in the middle fades into nothingness as we approach the inside of the two-pronged stimulus). Not even memory can now give us a full view of the alternatives.

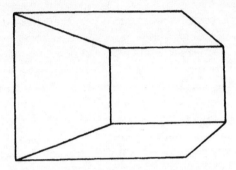

Every picture with only a modicum of perspective exhibits this phenomenon: we may decide to pay attention to the piece of paper on which the lines are drawn – but then there is no three-dimensional pattern; on the other hand we may decide to investigate the properties of this pattern, but then the surface of the paper disappears, or is integrated into what can only be called an illusion. There is no way of 'catching' the transition from the one to the other.[11] In all these cases the perceived image depends on 'mental sets' that can be changed at will, without the help of drugs, hypnosis, reconditioning. But mental sets may become frozen by illness, as a result of one's upbringing in a certain culture, or because of physiological determinants not in our control. (Not every change of language is accompanied by perceptual changes.) Our attitude towards other races, or towards people of a different cultural background, often depends on 'frozen' sets of the second kind: having learned to 'read'

---

11. See R.L. Gregory, *The Intelligent Eye*, London, 1970, Chapter 2. See also the distinction between eikon and phantasma in Plato, *Sophistes*, 235b8ff: 'This "appearing" or "seeming" without really "being" . . . all these expressions have always been and still are deeply involved in perplexity.' Plato talks about the distortions in statues of colossal size which were introduced to make them *appear* with the proper proportions. 'I cannot make use of an illusion and watch it,' says Gombrich in such cases, op. cit., p. 6.

faces in a standard way, we make standard judgements and are led astray.

An interesting example of physiologically determined sets leading to incommensurability is provided by the *development of human perception*. As has been suggested by Piaget and his school,[12] a child's perception proceeds through various stages before it reaches its relatively stable adult form. In one stage, objects seem to behave very much like after-images and are treated as such. The child follows the object with his eyes until it disappears; he does not make the slightest attempt to recover it, even if this should require but a minimal physical (or intellectual) effort, an effort, moreover, that is already within the child's reach. There is not even a tendency to search – and this is quite appropriate, 'conceptually' speaking. For it would indeed be nonsensical to 'look for' an after-image. Its 'concept' does not provide for such an operation.

The arrival of the concept, and of the perceptual image, of material objects, changes the situation quite dramatically. There occurs a drastic reorientation of behavioural patterns and, so one may conjecture, of thought. After-images, or things somewhat like them, still exist; but they are now difficult to find and must be discovered by special methods (the earlier visual world therefore literally *disappears*).[13] Such methods proceed from a new conceptual scheme (after-images occur in *humans*, they are not parts of the physical world) and cannot lead back to the exact phenomena of the previous stage. (These phenomena should therefore be called by a different name, such as 'pseudo-after-images' – a very interesting perceptual analogue to the transition from, say, Newtonian mechanics to special relativity: relativity, too, does not give us Newtonian facts, but relativistic analogues of Newtonian facts.) Neither after-images nor pseudo-after-images have a special position in the new world. For example, they are not treated as *evidence* on which the new notion of a material object is supposed to rest. Nor can they be used to *explain* this notion: after-images *arise together with it*, they depend on it, and are absent from the minds of those who do not yet recognize material objects; and pseudo-after-images *disappear* as soon as such recognition

---

12. J. Piaget, *The Construction of Reality in the Child*, New York, 1954, pp. 5ff.

13. This seems to be a general feature of the acquisition of new perceptual worlds: 'The older representations for the most part have to be suppressed rather than reformed,' writes Stratton in his epoch-making essay 'Vision without Inversion of the Retinal Image'. *The Psychological Review* IV, 1897, p. 471.

takes place. The perceptual field never contains after-images together with pseudo-after-images. It is to be admitted that every stage possesses a kind of observational 'basis' to which special attention is paid and from which a multitude of suggestions are received. However, this basis (a) *changes* from stage to stage, and (b) it is *part* of the conceptual apparatus of a given stage, not its one and only source of interpretation as some empiricists would like to make us believe.

Considering developments such as these, we may suspect that the family of concepts centring upon 'material object' and the family of concepts centring upon 'pseudo-after-image' are incommensurable in precisely the sense that is at issue here; these families cannot be used simultaneously and neither logical nor perceptual connections can be established between them.

Now is it reasonable to expect that conceptual and perceptual changes of this kind occur in childhood only? Should we welcome the fact, if it is a fact, that an adult is stuck with a stable perceptual world and an accompanying stable conceptual system, which he can modify in many ways but whose general outlines have forever become immobilized? Or is it not more realistic to assume that fundamental changes, entailing incommensurability, are still possible and that they should be encouraged lest we remain forever excluded from what might be a higher stage of knowledge and consciousness? Besides, the question of the mobility of the adult stage is at any rate an empirical question that must be attacked by *research*, and cannot be settled by methodological *fiat*.[14] The attempt to break through the boundaries of a given conceptual system is an essential part of such research (it also should be an essential part of any interesting life).

Such an attempt involves much more than a prolonged 'critical discussion',[15] as some relics of the enlightenment would have us believe. One must be able to *produce* and to *grasp* new perceptual and conceptual relations, including relations which are not immediately apparent (covert relations – see above) and *that* cannot be achieved by a critical discussion alone (see also above, Chapter 1 and 2). The orthodox

---

14. As Lakatos attempts to do: 'Falsification', p. 179, footnote 1: 'Incommensurable theories are neither inconsistent with each other, nor comparable for content. But we can *make* them, by a dictionary, inconsistent and their content comparable.'

15. Popper in *Criticism and the Growth of Knowledge*, p. 56.

accounts neglect the covert relations that contribute to their meaning, disregard perceptual changes and treat the rest in a rigidly standardized way so that any debate of unusual ideas is at once stopped by a series of routine responses. But now this whole array of responses is in doubt. Every concept that occurs in it is suspect, especially 'fundamental' concepts such as 'observation', 'test', and, of course, the concept 'theory' itself. And as regards the word 'truth', we can at this stage only say that it certainly has people in a tizzy, but has not achieved much else. The best way to proceed in such circumstances is to use examples which are outside the range of the routine responses. It is for this reason that I have decided to examine means of representation different from languages or theories and to develop my terminology in connection with them. More especially, I shall examine styles in painting and drawing. It will emerge that there are no 'neutral' objects which can be represented in any style, and which measure its closeness to 'reality'. The application to languages is obvious.

The 'archaic style' as defined by Emanuel Loewy in his work on ancient Greek art[16] has the following characteristics.

(1) The structure and the movement of the figures and of their parts are limited to a few typical schemes; (2) the individual forms are stylized, they tend to have a certain regularity and are 'executed with . . . precise abstraction';[17] (3) the representation of a form depends on the *contour* which may retain the value of an independent line or form the boundaries of a silhouette. 'The silhouettes could be given a number of postures: they could stand, march, row, drive, fight, die, lament. . . . But always their essential structure must be clear';[18] (4) *colour* appears in one shade only, and gradations of light and shadow are missing; (5) as a rule the figures

---

16. *Die Naturwiedergabe in der älteren Griechischen Kunst*, Rome, 1900, Chapter 1. Loewy uses 'archaic' as a *generic* term covering phenomena in Egyptian, Greek and Primitive Art, in the drawings of children and of untutored observers. In Greece his remarks apply to the *geometric style* (1000 to 700 BC) down to the *archaic period* (700 to 500 BC) which treats the human figure in greater detail and involves it in lively episodes. See also F. Matz, *Geschichte der Griechischen Kunst*, Vol. 1, 1950, as well as Beazly and Ashmole, *Greek Sculpture and Painting*, Cambridge, 1966, Chapters II and III.

17. Webster, *From Mycenae to Homer*, New York, 1964, p. 292. Webster regards this use of 'simple and clear patterns' in Greek geometric art as 'the forerunner of later developments in art (ultimately the invention of perspective), mathematics, and philosophy'.

18. Webster, op. cit., p. 205.

show their parts (and the larger episodes their elements) *in their most complete aspect* – even if this means awkwardness in composition, and 'a certain disregard of spatial relationships'. The parts are given their known value even when this conflicts with their seen relationship to the whole;[19] thus (6) with a few well-determined exceptions the figures which form a composition are arranged in such a way that *overlaps are avoided* and objects situated behind each other are presented as being side by side; (7) the *environment* of an action (mountains, clouds, trees, etc.) is either completely disregarded or it is omitted to a large extent. The action forms self-contained units of typical scenes (battles, funerals, etc.).[20]

These stylistic elements which are found, in various modifications, in the drawings of children, in the 'frontal' art of the Egyptians, in early Greek art, as well as among so-called Primitives, are explained by Loewy on the basis of psychological mechanisms: 'Side by side with the images which reality presents to the physical eye there exists an entirely different world of images which live or, better, come to life in our mind only and which, although suggested by reality, are totally transformed. Every primitive act of drawing ... tries to reproduce these images and them alone with the instinctive regularity of a psychical function.'[21] The archaic style changes as a result of 'numerous planned observations of nature which modify the pure mental images,'[22] initiate the development towards realism and thus start the history of art. *Natural*, physiological reasons are given for the archaic style and for its change.

Now it is not clear why it should be more 'natural' to copy memory images than images of perception which are better defined and more permanent.[23] We also find that realism often *precedes* more schematic forms of presentation. This is true of the Old Stone Age,[24] of Egyptian

---

19. Ibid., p. 207.

20. Beazly and Ashmole, op. cit., p. 3.

21. Loewy, op. cit., p. 4.

22. Ibid., p. 6.

23. The facts of perspective are noticed but they do not enter the pictorial presentation; this is seen from literary descriptions. See H. Schäfer, *Von Aegyptischer Kunst*, Wiesbaden, 1963, pp. 88ff, where the problem is further discussed.

24. See Paolo Graziosi, *Palaeolithic Art*, New York, 1960, and André Leroc-Gourhan, *Treasures of Prehistoric Art*, New York, 1967, both with excellent illustrations. These results were not known to Loewy: Cartailhac's 'Mea culpa d'un sceptique', for example, appeared only in 1902.

Art,[25] of Attic Geometric Art.[26] In all these cases the 'archaic style' is the result of a *conscious effort* (which may of course be aided, or hindered, by unconscious tendencies and physiological laws) rather than a natural reaction to internal deposits of external stimuli.[27] Instead of looking for the psychological *causes* of a 'style' we should therefore rather try to discover its *elements*, analyse their *function*, compare them with other phenomena of the same culture (literary style, sentence construction, grammar, ideology) and thus arrive at an outline of the underlying *world-view* including an account of the way in which this world-view influences perception, thought, argument, and of the limits it imposes on the roaming about of the imagination. We shall see that such an analysis of outlines provides a better understanding of the process of conceptual change than either a naturalistic account which recognizes only one 'reality' and orders artworks by their closeness to it, or trite slogans such as 'a critical discussion and a comparison of . . . various frameworks is always possible'.[28] Of course, *some* kind of comparison is *always* possible (for example, one physical theory may sound more melodious when read aloud to the accompaniment of a guitar than another physical theory). But lay down *specific* rules for the process of comparison, such as the rules of logic as applied to the relation of content classes, or some simple rules of perspective and you will find exceptions, undue restrictions, and

---

25. See the change in the presentation of animals in the course of the transition from predynastic times to the First Dynasty. The Berlin lion (Berlin, Staatliches Museum, Nr. 22440) is wild, threatening, quite different in expression and execution from the majestic animal of the Second and Third Dynasties. The latter seems to be more a representation of the *concept* lion than of any individual lion. See also the difference between the falcon on the victory tablet of King Narmer (backside) and on the burial stone of King Wadji (Djet) of the First Dynasty. 'Everywhere one advanced to pure clarity, the forms were strengthened and made simple,' Schäfer, op. cit., pp. 12ff, especially p. 15, where further details are given.

26. 'Attic geometric art should not be called primitive, although it has not the kind of photographic realism which literary scholars seem to demand in painting. It is a highly sophisticated art with its own conventions which serve its own purposes. As with the shapes and the ornamentation, a revolution separates it from late Mycenaean painting. In this revolution figures were reduced to their minimum silhouettes, and out of these minimum silhouettes the new art was built up.' Webster, op. cit., p. 205.

27. This thesis is further supported by the observation that so-called Primitives often turn their back to the objects they want to draw; Schäfer, op. cit., p. 102, after Conze.

28. Popper, in *Criticism*, etc., p. 56.

you will be forced to talk your way out of trouble at every turn. It is much more interesting and instructive to examine what kinds of things can be said (represented) and what kinds of things cannot be said (represented) *if the comparison has to take place within a certain specified and historically well-entrenched framework.* For such an examination we must go beyond generalities and study frameworks in detail. I start with an account of some examples of the archaic style.

The human figure shows the following characteristic: 'the men are very tall and thin, the trunk of a triangle tapering to the waist, the head of a knob with a mere excrescence for a face: towards the end of the style the head is lit up – the head knob is drawn in outline, and a dot signifies the eye'.[29] All, or almost all, parts are shown in profile and they are strung together like the limbs of a puppet or a rag doll. They are not 'integrated' to form an organic whole. This 'additive' feature of the archaic style becomes very clear from the treatment of the eye. The eye does not participate in the actions of the body, it does not guide the body or establish contact with the surrounding situation, it does not 'look'. It is added on to the profile head like part of a notation, as if the artist wanted to say: 'and beside all these other things such as legs, arms, feet, a man has also eyes, they are in the head, one on each side'. Similarly, special states of the body (alive, dead, sick) are not indicated by a special arrangement of its parts, but by putting the same standard body into various standard *positions.* Thus the body of the dead man in a funeral carriage is articulated in exactly the same way as that of a standing man, but it is rotated through 90 degrees and inserted in the space between the bottom of the shroud and the top of the bier.[30] Being shaped like the body of a live man it is *in addition* put into the death position. Another instance is the picture of a kid half swallowed by a lion.[31] The lion looks ferocious, the kid looks peaceful, and the act of swallowing is simply *tacked on* to the presentation of what a lion *is* and what a kid *is.* (We have what is called a *paratactic aggregate*: the elements of such an aggregate are all given equal importance, the only relation

---

29. Beazly and Ashmole, op. cit., p. 3.

30. Webster, op. cit., p. 204: 'The painter feels the need to say that he has two arms, two legs, and a manly chest.'

31. R. Hampl, *Die Gleichnisse Homers und die Bildkunst seiner Zeit*, Tübingen, 1952.

between them is sequential, there is no hierarchy, no part is presented as being subordinate to and determined by others.) The picture *reads*: ferocious lion, peaceful kid, swallowing of kid by lion.

The need to show every essential part of a situation often leads to a separation of parts which are actually in contact. The picture becomes a list. Thus a charioteer standing in a carriage is shown as standing above the floor (which is presented in its fullest view) and unencumbered by the rails so that his feet, the floor, the rails can all be clearly seen. No trouble arises if we regard the painting as a *visual catalogue* of the parts of an event rather than as an illusory rendering of the event itself (no trouble arises when we *say*: his *feet* touched the *floor* which is *rectangular*, and he was surrounded by a *railing*. . .)[32] But such an interpretation must be *learned*, it cannot be simply read off the picture.

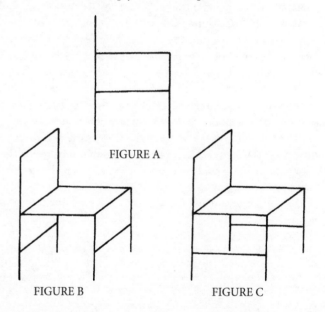

FIGURE A

FIGURE B   FIGURE C

The amount of learning needed may be considerable. Some Egyptian drawings and paintings can be decoded only with the help of either the

---

32. 'All geometric pictures of chariots show at least one of these distortions.' Webster, op. cit., p. 204. Late Mycenaean pottery, on the other hand, has the legs of the occupants concealed by the side.

represented object itself or with the help of three-dimensional copies of it (statuary in the case of humans, animals, etc.). Using such information we learn that the chair in Figure A represents the object of Figure C and not the object of Figure B and that it must be read: 'chair with backrest and four legs, legs connected by support' where it is understood that the front legs are connected with the back legs and not with each other.[33] The interpretation of groups is complicated and some cases are not yet understood.[34]

(Being able to 'read' a certain style also includes knowledge of what features are *irrelevant*. Not every feature of an archaic list has representational value, just as not every feature of a written sentence plays a role in articulating its content. This was overlooked by the Greeks, who started inquiring into the reasons for the 'dignified postures' of Egyptian statues [already Plato commented on this]. Such a question 'might have struck an Egyptian artist as it would strike us if someone inquired about the age or the mood of the king on the chessboard'.[35])

So far a brief account of some peculiarities of the 'archaic' style. A style can be described and analysed in various ways. The descriptions given so far paid attention to *formal features*: the archaic style provides *visible lists* whose parts are arranged in roughly the same way in which they occur in 'nature' except when such an arrangement is liable to hide important elements. All parts are on the same level, we are supposed to 'read' the lists rather than 'see' them as illusory accounts of the situation.[36] The lists are not organized in any way except sequentially, that is, the shape of

---

33. Schäfer, op. cit., p. 123.

34. Ibid., pp. 223ff.

35. Gombrich, op. cit., p. 134, with literature.

36. 'We come closer to the factual content of frontal [*geradvorstelliger*] drawings of objects, if we start by *reading off* their partial contents in the form of narrative declarative sentences. The frontal mode of representation gives us a "visual concept" [*Sehbegriff*] of the thing (the situation) represented.' Schäfer, op. cit., p. 118. See also Webster, op. cit., p. 202, about the 'narrative' and 'explanatory' character of Mycenaean and geometric art. But see H.A. Groenewegen-Frankfort, *Arrest and Movement*, London, 1951, pp. 33f: the scenes from daily life on the walls of Egyptian tombs 'should be "read": harvesting entails ploughing, sowing, and reaping; care of cattle entails fording of streams and milking . . . the sequence of scenes is purely conceptual, not narrative, nor is the writing which occurs with the scenes dramatic in character. The signs, remarks, names, songs and explanations, which illuminate the action . . . do not link events or explain their development; they are typical sayings belonging to typical situations.'

an element does not depend on the presence of other elements (adding a lion and the act of swallowing does not make the kid look unhappy; adding the process of dying does not make a man look weak). Archaic pictures are *paratactic aggregates*, not hypotactic systems. The elements of the aggregate may be physical parts such as heads, arms, wheels; they may be states of affairs such as the fact that a body is dead; they may be actions, such as the action of swallowing.

Instead of describing the formal features of a style, we may describe the *ontological features* of a world that consists of the elements represented in the style, arranged in the appropriate way, and we may also describe the *impression* such a world makes upon the viewer. This is the procedure of the art critic who loves to dwell on the peculiar behaviour of the characters which the artist puts on his canvas and on the 'internal life' the behaviour seems to indicate. Thus G.M.S. Hanfmann[37] writes on the archaic figure: 'No matter how animated and agile archaic heroes may be, they do not appear to move by their own will. Their gestures are explanatory formulae imposed upon the actors from without in order to explain what sort of action is going on. And the crucial obstacle to the convincing portrayal of inner life was the curiously detached character of the archaic eye. It shows that a person is alive, but it cannot adjust itself to the demands of a specific situation. Even when the archaic artist succeeds in denoting a humorous or tragic mood, these factors of externalized gesture and detached glance recall the exaggerated animation of a puppet play.'

An ontological description frequently adds just verbiage to the formal analysis; it is nothing but an exercise in 'sensitivity' and cuteness. However, we must not disregard the possibility that a particular style *gives a precise account of the world as it is perceived by the artist and his contemporaries* and that every formal feature corresponds to (hidden or explicit) assumptions inherent in the underlying cosmology. (In the case of the 'archaic' style we must not disregard the possibility that humans then actually *felt* themselves to be what we today would call puppets guided by outside forces and that they *saw* and *treated* others accordingly.) Such a *realistic interpretation* of styles would be in line with Whorf's thesis that in addition to being instruments for *describing* events (which may

---

37. 'Narration in Greek Art', *American Journal of Archaeology*, Vol. 61, January 1957, p. 74.

have other features, not covered by any description) languages are also *shapers* of events (so that there is a linguistic limit to what can be said in a given language, and this limit coincides with the limits of the thing itself) but it would go beyond it by including non-linguistic means of representation.[38] The realistic interpretation is very plausible. But it must not be taken for granted.[39]

It must not be taken for granted, for there are technical failures, special purposes (caricature) which may change a style without changing the cosmology. We must also remember that humans have roughly the same neurophysiological equipment, so that perception cannot be bent in any direction one chooses.[40] And in some cases we can indeed show that deviations from a 'faithful rendering of nature' occur in the presence of a detailed knowledge of the object and side by side with more 'realistic' presentations: the workshop of the sculptor Thutmosis in Tel al-Amarna (the ancient Achet-Aton) contains masks directly taken from live models with all the details of the formation of the head (indentations) and of the face intact, as well as heads developed from such masks. Some of these heads preserve the details, others eliminate them and replace them by simple forms. An extreme example is the completely smooth head of an Egyptian man. It proves that 'at least some artists remained consciously independent of nature'.[41] During the reign of Amenophis IV (BC 1364–1347) the mode of representation was changed twice; the first change, towards a more realistic style, occurred merely four years after his ascension to the throne, which shows that the technical ability for realism existed, was ready for use, but was intentionally left undeveloped. *An inference from style (or language) to cosmology and modes of perception therefore needs special argument: it cannot be made as a matter of course.* (A similar remark applies to any inference from popular theories in science, such as the theory of relativity, or the idea of the motion of the earth, to cosmology and modes of perception.)

---

38. See footnote 1 and text of the present chapter.

39. For a sketch of the problems that arise in the case of *physical theories,* see my 'Reply to Criticism', *Boston Studies in the Philosophy of Science,* Vol. 2, 1965, sections 5–8, and especially the list of problems on p. 234. Hanson, Popper and others take it for granted that realism is correct.

40. It may be different with drug-induced states, especially when they are made part of a systematic course of education.

41. Schäfer, op. cit., p. 63.

The argument (which can never be conclusive) consists in pointing to characteristic features in distant fields. If the idiosyncrasies of a particular style of painting are found also in statuary, in the grammar of contemporary languages (and here especially in covert classifications which cannot be easily twisted around); if it can be shown that these languages are spoken by artists and by the common folk alike; if there are philosophical principles formulated in the languages which declare the idiosyncrasies to be features of the world and not just artifacts and which try to account for their origin; if man and nature have these features not only in paintings, but also in poetry, in popular sayings, in common law; if the idea that the features are parts of normal perception is not contradicted by anything we know from physiology, or from the psychology of perception; if later thinkers attack the idiosyncrasies as 'errors' resulting from an ignorance of the 'true way', then we may assume that we are not just dealing with technical failures and particular purposes, *but with a coherent way of life*, and we may expect that people involved in this way of life see the world in the same way in which we now see their pictures. It seems that all these conditions are satisfied in archaic Greece: the formal structure and the ideology of the *Greek epic* as reconstructed both from the text and from later references to it repeat all the peculiarities of the later geometric and the early archaic style.[42]

To start with, about nine-tenths of the Homeric epics consist of *formulae* which are prefabricated phrases extending in length from a single word or two to several complete lines and which are repeated at appropriate places.[43] One-fifth of the poems consist of lines wholly repeated from one place to another; in 28,000 Homeric lines there are about 25,000 repeated phrases. Repetitions occur already in Mycenaean court poetry and they can be traced to the poetry of eastern courts: 'Titles of gods, kings, and men must be given correctly, and in a courtly world the principle of correct expression may be extended further. Royal correspondence is highly formal, and this formality is extended beyond the messenger

---

42. Webster, op. cit., pp. 294ff.

43. In the 20th century the role of formulae was described and tested by Milman Parry, *L'Epithète traditionelle chez Homère*, Paris, 1928; *Harvard Studies in Classical Philology*, Vols. 41, 1930, 43, 1932. For a brief account see D.I. Page, *History and the Homeric Iliad*, Berkeley, 1966, Chapter VI, as well as G.S. Kirk, *Homer and the Epic*, Cambridge, 1965, Part I.

scenes of poetry to the formulae used for introducing speeches. Similarly, operations are reported in the terms of the operation order, whether the operation order itself is given or not, and this technique is extended to other descriptions, which have no such operation orders behind them. These compulsions all derive ultimately from the court of the king, and it is reasonable to suppose that the court in turn enjoyed such formality in poetry.'[44] The conditions of (Sumerian, Babylonian, Hurrian, Hethitic, Phoenician, Mycenaean) courts also explain the occurrence of standardized elements of *content* (typical scenes; the king and the nobles in war and peace; furniture; description of beautiful things) which, moving from city to city, and even across national boundaries are repeated, and adapted to local circumstances.

The slowly arising combination of constant and variable elements that is the result of numerous adaptations of this kind was utilized by the illiterate poets of the 'Dark Age' of Greece who developed a language and forms of expression that best serve the requirements of *oral composition*. The requirement of *memory* demanded that there be ready-made descriptions of events that can be used by a poet who composes in his mind, and without the aid of writing. The requirement of *metre* demanded that the basic descriptive phrases be fit for use in the various parts of the line the poet is about to complete: 'Unlike the poet who writes out his lines . . . [the oral poet] cannot think without hurry about his next word, nor change what he has made nor, before going on, read over what he has just written. . . . He must have for his use word groups all made to fit his verse.'[45] *Economy* demanded that, given a situation and a certain metrical constraint (beginning, middle or end of a line), there be only one way of continuing the narration – and this demand is satisfied to a surprising extent: 'All the chief characters of the *Iliad* and the *Odyssey*, if their names can be fitted into the last half of the verse along with an epithet, have a noun-epithet formula in the nominative, beginning with a simple consonant, which fills the verse between the trochaic caesura of the third foot and the verse end: for instance, πολύτλαφ διοφ Ὀδυσσευφ. In a list of thirty-seven characters who have formulae of this type, which includes all those having any importance in the poems, there are only three names which have a second formula which could

---

44. Webster, op. cit., pp. 75f.
45. M. Parry, *Harvard Studies Classical Philology*, 41, 1930, p. 77.

replace the first.'[46] 'If you take in the five grammatic cases the singular of all the noun-epithet formulae used for Achilles, you will find that you have forty-five different formulae of which none has, in the same case, the same metrical value.'[47] Being provided for in this manner, the Homeric poet 'has no interest in originality of expression, or in variety. He uses or adapts inherited formulae.'[48] He does not have a 'choice, do[es] not even think in terms of choice; for a given part of the line, whatever declension case was needed, and whatever the subject matter might be, the formular vocabulary supplied at once a combination of words ready-made'.[49]

Using the formulae the Homeric poet gives an account of *typical scenes* in which objects are occasionally described by 'adding the parts on in a *string of words* in apposition'.[50] Ideas we would today regard as being logically subordinate to others are stated in separate, grammatically co-ordinate propositions. Example (*Iliad*, 9.556ff): Meleagros 'lay by his wedded wife, fair Cleopatra, daughter of fair-ankled Marpessa, daughter of Euenos, and of Ides, who was the strongest of men on earth at that time – and he against lord Phoebus Apollo took up his bow for the sake of the fair-ankled maid: her then in their halls did her father and lady mother call by the name of Alkyon because . . .' and so on, for ten more lines and two or three more major themes before a major stop. This *paratactic* feature of Homeric poetry which parallels the absence of elaborate systems of subordinate clauses in early Greek[51] also makes it clear why Aphrodite

---

46. Ibid., pp. 86f.
47. Ibid., p. 89.
48. Page, op. cit., p. 230.
49. Ibid., p. 242.
50. Webster, op. cit., pp. 99f; my italics.
51. See Raphael Kühner, *Ausführliche Grammatik der Griechischen Sprache*, 2. Teil, reprinted Darmstadt, 1966. In the 20th century such a paratactic or 'simultanistic' way of presentation was used by the early expressionists, for example by Jacob von Hoddis in his poem *Weltende*:

> Dem Bürger fliegt vom spitzen Kopf der Hut,
> In allen Lüften hallt es wie Geschrei.
> Dachdecker stürzen ab und gehn entzwei,
> Und an den Küsten – liest man – steigt die Flut.
> Der Sturm ist da, die wilden Meere hupfen
> An Land, um dicke Dämme zu zerdrücken.
> Die meisten Menschen haben einen Schnupfen.
> Die Eisenbahnen fallen von den Brucken.

is called 'sweetly laughing' when in fact she complains tearfully (*Iliad*, 5.375), or why Achilles is called 'swift footed' when he is sitting talking to Priam (*Iliad*, 24.559). Just as in late geometric pottery (in the 'archaic' style of Loewy) a dead body is a live body brought into the position of death (see above, text to footnote 30) or an eaten kid a live and *peaceful kid* brought into the appropriate relation to the mouth of a ferocious lion, in the very same way Aphrodite complaining is simply Aphrodite – and that is the laughing goddess – *inserted* into the situation of complaining in which she participates only externally, without changing her nature.

The *additive treatment* of events becomes very clear in the case of (human) motion. In *Iliad*, 22.298, Achilles drags Hector along in the dust 'and dust arose around him that was dragged, and his dark hair flowed loose on either side, and in the dust *lay* his once fair head' – that is, the *process* of dragging contains the *state* of lying as an independent part which together with other such parts constitutes the motion.[52] Speaking more abstractly, we might say that for the poet 'time is composed of moments'.[53] Many of the similes assume that the parts of a complex entity have a life of their own and can be separated with ease. Geometrical man is a visible list of parts and positions; Homeric man is put together from

---

Von Hoddis claims Homer as a precursor, explaining that simultaneity was used by Homer not in order to make an event more transparent but in order to create a feeling of immeasurable spaciousness. When Homer describes a battle and compares the noise of the weapons with the beat of a woodcutter, he merely wants to show that while there is battle there is also the quietness of woods, interrupted only by the work of the woodcutter. Catastrophe cannot be thought without simultaneously thinking of some utterly unimportant event. The Great is mixed up with the Small, the Important with the Trivial. (For the report see J.R. Becher in *Expressionismus*, ed. P. Raabe, Olten and Freiburg, 1965, pp. 50ff; this short article also contains a description of the tremendous impression von Hoddis' eight-liner made when it first came out in 1911.) One cannot infer that the same impression was created in the listener of the Homeric singers who did not possess a complex and romanticizing medium that had deteriorated into tearful sentimentality as a background for comparison.

52. See Gebhard Kurz, *Darstellungsformen menschlicher Bewegung in der Ilias*, Heidelberg, 1966, p. 50.

53. This is the theory ascribed to Zeno by Aristotle, *Physics*, 239b, 31. The theory comes forth most clearly in the argument of the arrow: 'The arrow at flight is at rest. For, if everything is at rest when it occupies a space equal to itself, and what is in flight at any given moment always occupies a space equal to itself, it cannot move' (after *Physics*, 239b). We cannot say that the theory was held by Zeno himself, but we may conjecture that it played a role in Zeno's time.

limbs, surfaces, connections which are isolated by comparing them with inanimate objects of precisely defined shape: the trunk of Hippolochos rolls through the battle field like a *log* after Agamemnon has cut off his arms and his head (*Iliad*, 11.146 –ολμοϑ, round stone of cylindrical shape), the body of Hector spins like a top (*Iliad*, 14.412), the head of Gorgythion drops to one side 'like a *garden poppy* being heavy with fruit and the showers of spring' (*Iliad*, 8.302);[54] and so on. Also, the formulae of the epic, especially the noun-epithet combinations, are frequently used not according to content but according to metrical convenience: 'Zeus changes from counsellor to storm-mountain god to paternal god *not* in connection with what he is doing, but at the dictates of metre. He is not *nephelegerata Zeus* when he is gathering clouds, but when he is filling the metrical unit, $\cup\cup - \cup\cup - -$',[55] just as the geometrical artist may distort spatial relations – introduce contact where none exists and break it where it occurs – in order to tell the visual story in his own particular way. Thus the poet repeats the formal features used by the geometric and the early archaic artists. Neither seems to be aware of an 'underlying substance' that keeps the objects together and shapes their parts so that they reflect the 'higher unity' to which they belong.

Nor is such a 'higher unity' found in the concepts of the language. For example, there is no expression that could be used to describe the human body as a single entity.[56] *Soma* is the corpse, *demas* is accusative of specification, it means 'in structure', or 'as regards shape', reference to *limbs* occurs where we today speak of the body (γυια, limbs as moved by the joints; μέλεα, limbs in their bodily strength; λέλυντο γυια, his whole body trembled; ιδϑ εχ μελέων ετ11εν, his body was filled with strength). All we get is a puppet put together from more or less articulated parts.

The puppet does not have a soul in our sense. The 'body' is an aggregate of limbs, trunk, motion; the 'soul' is an aggregate of 'mental' events which

---

54. Kurz, op. cit.

55. R. Lattimore, *The Iliad of Homer*, Chicago, 1951, pp. 39f.

56. For the following see B. Snell, *The Discovery of the Mind*, Harper Torchbooks, 1960, Chapter 1. Snell's views have been criticized but seem to survive the criticism. See the report in F. Krafft, *Vergleichende Untersuchungen zu Homer und Hesiod, Hypomnemata*, Heft 6, Göttingen, 1963, pp. 25ff. In his *Gesammelte Schriften*, Göttingen, 1966, p. 18, Snell also argues that 'in Homer we never find a personal decision, a conscious choice made by an acting human being. A human being who is faced with various possibilities never thinks: "now it depends on me, it depends on what I decide to do".'

are not necessarily private and which may belong to a different individual altogether. 'Never does Homer in his description of ideas or emotions go beyond a purely spatial, or quantitative definition; never does he attempt to sound their special, non-physical nature.'[57] Actions are initiated not by an 'autonomous I', but by further actions, events, occurrences, including divine interference. And this is precisely how mental events are *experienced*.[58] Dreams, unusual psychological feats such as sudden remembering, sudden acts of recognition, sudden increase of vital energy, during battle, during a strenuous escape, sudden fits of anger are not only *explained* by reference to gods and demons, they are also *felt* as such. Agamemnon's dream 'listened to his [Zeus'] words and descended' (*Iliad*, 2.16) – the *dream* descends, not a figure in it – 'and it stood then beside his [Agamemnon's] head in the likeness of Nestor' (*Iliad*, 2.20). One does not *have* a dream (a dream is not a 'subjective' event), one *sees* it (it is an 'objective' event) and one also sees how it approaches and moves away.[59] Sudden anger, fits of strength are described *and felt* to be divine acts:[60] 'Zeus builds up and Zeus diminishes strength in man the way he pleases, since his power is beyond all others' (*Iliad*, 20.241) is not just an objective description (that may be extended to include the behaviour of animals), it also expresses the *feeling* that the change has entered from the outside, that one has been 'filled . . . with strong courage' (*Iliad*, 13.60). Today such events are either forgotten or regarded as purely

---

57. Snell, *Gesammelte Schriften*, p. 18.

58. See Dodds, *The Greeks and the Irrational*, Boston, 1957, Chapter 1.

59. With some effort this experience can be repeated even today. Step 1: lie down, close your eyes, and attend to your hypnagogic hallucinations. Step 2: permit the hallucinations to proceed on their own and according to their own tendencies. They will then change from events in front of the eyes into events that gradually surround the viewer but without yet making him an active participant of an action in a three-dimensional dream-space. Step 3: switch over from *viewing* the hallucinatory event to *being part* of a complex of real events which act on the viewer and can be acted upon by him. Step 3 can be reversed either by the act of an almost non-existent will or by an outside noise. The three-dimensional scenery becomes two-dimensional, runs together into an area in front of the eyes, and moves away. It would be interesting to see how such *formal* elements change from culture to culture.

60. Today we say that somebody is 'overcome' by emotions and he may feel his anger as an alien thing that invades him against his will. The daemonic ontology of the Greeks contains objective terminology for describing this feature of our emotions *and thereby stabilizes it*.

accidental.[61] 'But for Homer, or for early thought in general, there is no such thing as accident.'[62] Every event is accounted for. This makes the events clearer, strengthens their objective features, moulds them into the shape of known gods and demons and thus turns them into powerful evidence for the divine apparatus that is used for explaining them: 'The gods are present. To recognize this as a given fact for the Greeks is the first condition for comprehending their religion and their culture. Our knowledge of their presence rests upon an (inner or outer) experience of either the Gods themselves or of an action of the Gods.'[63]

To sum up: the archaic world is much less compact than the world that surrounds us, and it is also experienced as being less compact. Archaic man lacks 'physical' unity, his 'body' consists of a multitude of parts, limbs, surfaces, connections; and he lacks 'mental' unity, his 'mind' is composed of a variety of events, some of them not even 'mental' in our sense, which either inhabit the body-puppet as additional constituents or are brought into it from the outside. Events are not *shaped* by the individual, they are complex arrangements of parts into which the body-puppet is *inserted* at the appropriate place.[64] This is the world-view that emerges from an

---

61. Psychoanalysis and related ideologies now again contribute to making such events part of a wider context and thereby lend them substantiality.

62. Dodds, op. cit., p. 6.

63. Wilamowitz-Moellendorf, *Der Glaube der Hellenen*, 1, 1955, p. 17. Our conceptions of the world subdivide an otherwise uniform material and create differences in perceived brightness where objective brightness has no gradient. The same process is responsible for the ordering of the rather chaotic impressions of our inner life, leading to an (inner) perception of divine interference, and it may even introduce daemons, gods, sprites into the domain of outer perceptions. At any rate – there is a sufficient number of daemonic experiences not to reject this conjecture out of hand.

64. This means that success is not the result of an effort on the part of the individual but the fortunate fitting together of circumstances. This shows itself even in words like πάττειν, which seem to designate *activities*. In Homer such words emphasize not so much the effect of the agent as the fact that the result comes about in the right way, that the process that brings it about does not encounter too many disturbances; it fits into the other processes that surround it (in the Attic dialect επάττω still means 'I am doing well'). Similarly τεύχειν emphasizes not so much a personal achievement as the fact that things go well, that they fit into their surroundings. The same is true of the acquisition of knowledge. 'Odysseus has seen a lot and experienced much, moreover, he is the πολυμήχανοφ who can always help himself in new ways, and, finally, he is the man who listens to his goddess Athena. The part of knowledge that is based on seeing is not really the result of his own activity and research, it rather happened to him

analysis of the *formal* features of 'archaic' art and Homeric poetry, taken in conjunction with an analysis of the *concepts* which the Homeric poet used for describing what he perceives. Its main features are *experienced* by the individuals using the concepts. *These individuals live indeed in the same kind of world that is constructed by their artists.*

Further evidence for the conjecture can be obtained from an examination of 'meta-attitudes' such as general religious attitudes and 'theories' of (attitudes to) knowledge.

For the lack of compactness just described reappears in the field of ideology. There is a *tolerance* in religious matters which later generations found morally and theoretically unacceptable and which even today is regarded as a manifestation of frivolous and simple minds.[65] Archaic man is a religious eclectic, he does not object to foreign gods and myths, he adds them to the existing furniture of the world without any attempt at synthesis, or a removal of contradictions. There are no priests, there is no dogma, there are no categorical statements about the gods, humans, the world.[66] (This tolerance can still be found with the Ionian philosophers of nature who develop their ideas side by side with myth without trying to eliminate the latter.) There is no religious 'morality' in our sense, nor are the gods abstract embodiments of eternal principles.[67] This they became later, during the archaic age, and as a result they 'lost [their] humanity. Hence Olympianism in its moralized form tended to become a religion of fear, a tendency which is reflected in the religious vocabulary. There is no word for "god-fearing" in the *Iliad*.'[68] This is how life was dehumanized by what some people are pleased to call 'moral progress' or 'scientific progress'.

---

while he was driven around by external circumstances. He is very different from Solon who, as Herodotus tells us, was the first to travel for theoretical reasons, because he was interested in research. In Odysseus the knowledge of many things is strangely separated from his activity in the field of the πίσασθαι: this activity is restricted to finding means for reaching a certain aim, in order to save his life and the life of his associates.' B. Snell, *Die alten Griechen und Wir*, Göttingen, 1962, p. 48. In this place also a more detailed analysis of pertinent terms. See also footnote 56 on the apparent non-existence of personal decisions.

65. Example: F. Schachermayer, *Die frühe Klassik der Griechen*, Stuttgart, 1966.
66. See Wilamowitz–Moellendorf, op. cit.
67. M.P. Nilsson, *A History of Greek Religion*, Oxford, 1949, p. 152.
68. Dodds, op. cit., p. 35.

Similar remarks apply to the 'theory of knowledge' that is implicit in this early world-view. The Muses in *Iliad*, 2.284ff, have knowledge because they are *close* to things – they do not have to rely on rumours – and because they know all the *many* things that are of interest to the writer, one after the other. 'Quantity, not intensity is Homer's standard of judgement' and of knowledge,[69] as becomes clear from such words as πολ ὑψ#τιων and πολ ὑμητιφ, 'much pondering' and 'much thinking', as well as from later criticisms such as 'Learning of many things [πολυμαθίη] does not teach intelligence'.[70] An interest in, and a wish to understand, *many amazing things* (such as earthquakes, eclipses of the sun and the moon, the paradoxical rising and falling of the Nile), each of them explained in its own particular way and *without* the use of universal principles, persists in the coastal descriptions of the 8th and 7th (and later) centuries (which simply *enumerate* the tribes, tribal habits, and coastal formations that are successively met during the journey), and even a thinker such as Thales is satisfied with making many interesting observations and providing many explanations without trying to tie them together in a system.[71] (The first thinker to construct a 'system' was Anaximander, who followed Hesiod.) *Knowledge* so conceived is not obtained by trying to grasp an essence behind the reports of the senses, but by (1) putting the observer in the right position relative to the object (process, aggregate), by inserting him into the appropriate place in the complex pattern that constitutes the world, and (2) by adding up the elements which are noted under these circumstances. It is the result of a complex survey carried out from suitable vantage points. One may doubt a vague report, or a fifth-hand account, but it is not possible to doubt what one can clearly see with one's own eyes. The *object* depicted or described is the proper arrangements of the elements which may include foreshortenings and other perspectoid

---

69. Snell, *The Discovery of the Mind*, p. 18.

70. Heraclitus, after Diogenes Laertius, IX, 1.

71. The idea that Thales used a principle expressing an underlying unity of natural phenomena and that he identified this principle with water is first found in Aristotle, *Metaphysics*, 983b6–12 and 26ff. A closer look at this and other passages and consultation of Herodotus suggests that he still belongs to the group of those thinkers who deal with numerous extraordinary phenomena, without tying them together in a system. See the vivid presentation in F. Krafft, *Geschichte der Naturwissenschaften*, I, Freiburg, 1971, Chapter 3.

phenomena.[72] The fact that an oar looks broken in water lacks here the sceptical force it assumes in another ideology.[73] Just as Achilles sitting does not make us doubt that he is swift-footed – as a matter of fact, we would start doubting his swiftness if it turned out that he is in principle incapable of sitting – in the very same way the bent oar does not make us doubt that it is perfectly straight in air – as a matter of fact, we would start doubting its straightness if it did not look bent in water.[74] The bent oar is not an *aspect* that denies what another *aspect* says about the *nature* of the oar, it is a particular *part* (situation) of the real oar that is not only *compatible* with its straightness, but that demands it: the objects of knowledge are as additive as the visible lists of the archaic artist and the situations described by the archaic poet.

Nor is there any uniform conception of knowledge.[75] A great variety of words is used for expressing what we today regard as different forms of knowledge, or as different ways of acquiring knowledge. σοφία[76] means expertise in a certain profession (carpenter, singer, general, physician, charioteer, wrestler) including the arts (where it praises the artist not as an outstanding creator but as a master of his craft); εδεναι, literally 'having seen', refers to knowledge gained from inspection; συνίημι, especially in the

---

72. Perspectoid phenomena are sometimes treated as if they were special properties of the objects depicted. For example, a container of the Old Kingdom (Ancient Egypt) has an indentation on top, indicating perspective, but the indentation is presented as a feature of the object itself, Schäfer, op. cit., p. 266. Some Greek artists try to find situations where perspective does not need to be considered. Thus the peculiarity of the so-called red-figure style that arises in about 530 BC 'does not so much consist in the fact that foreshortenings are drawn, but in the new and highly varied ways to circumvent them', E. Pfuhl, *Malerei und Zeichung der Griechen*, Vol. 1, Munich, 1923, p. 378.

73. See the discussion in Chapter 1 and A.J. Ayer's *Foundations of Empirical Knowledge*. The example was familiar in antiquity.

74. This is also the way in which J.L. Austin takes care of the case. See *Sense and Sensibilia*, New York, 1962. It is clear that problems such as the 'problem of the existence of theoretical entities' cannot arise under these circumstances either. All these problems are *created* by the new approach that superseded the additive ideology of archaic and pre-archaic times.

75. B. Snell, *Die Ausdrücke für den Begriff des Wissens in der vorplatonischen Philosophie*, Berlin, 1924. A short account is given in Snell, *Die alten Griechen und wir*, pp. 41ff. See also von Fritz, *Philosophie und sprachlicher Ausdruck bei Demokrit, Plato, und Aristoteles*, Leipzig–Paris–London, 1938.

76. Only occurrence in Homer, *Iliad*, 15, 42, concerning the σοφία of a carpenter (an 'expert carpenter' translates Lattimore).

Iliad, though often translated as 'listening' or 'understanding', is stronger – it contains the idea of following and obeying, one absorbs something and acts in accordance with it (hearing may play an important role). And so on. Many of these expressions entail a receptive attitude on the part of the knower; he, as it were, acts out the behaviour of the things around him, he follows them,[77] he acts as befits an entity that is inserted at the place he occupies.

To repeat and to conclude: the modes of representation used during the early archaic period in Greece are not just reflections of incompetence or of special artistic interests, they give a faithful account of what are felt, seen, thought to be fundamental features of the world of archaic man. This world is an open world. Its elements are not formed or held together by an 'underlying substance', they are not appearances from which this substance may be inferred with difficulty. They occasionally coalesce to form assemblages. The relation of a single element to the assemblage to which it belongs is like the relation of a part to an aggregate of parts and not like the relation of a part to an overpowering whole. The particular aggregate called 'man' is visited, and occasionally inhabited, by 'mental events'. Such events may reside in him, they may also enter from the outside. Like every other object man is an exchange station of influences rather than a unique source of action, an 'I' (Descartes' 'cogito' has no point of attack in this world, and his argument cannot even start). There is a great similarity between this view and Mach's cosmology except that the elements of the archaic world are recognizable physical and mental shapes and events while Mach's elements are more abstract, they are as yet unknown *aims* of research, not its *object*. In sum, the representational units of the archaic world-view admit of a realistic interpretation, they express a coherent ontology, and Whorf's observations apply.

At this point I interrupt my argument in order to make some comments which connect the preceding observations with the problems of scientific method.

1. It may be objected that foreshortenings and other indications of perspective are such obvious features of our perceptual world that they cannot have been absent from the perceptual world of the Ancients. The archaic manner of presentation is therefore incomplete, and its realistic interpretation incorrect.

---

77. See Snell, *Ausdrücke*, p. 50.

Reply: Foreshortenings are not an obvious feature of our perceptual world unless special attention is drawn to them (in an age of photography and film this is rather frequently the case). Unless we are professional photographers, film-makers, painters we perceive *things*, not *aspects*. Moving swiftly among complex objects we notice much less change than a perception of aspects would permit. Aspects, foreshortenings, if they enter our consciousness at all, are usually suppressed just as after-images are suppressed when the appropriate stage of perceptual development is completed[78] and they are noticed in special situations only.[79] In ancient Greece such special situations arose in the theatre, for the first-row viewers of the impressive productions of Aeschylus and Agatharchos, and there is indeed a school that ascribes to the theatre a decisive influence on the development of perspective.[80] Besides, why should the perceptual world of the ancient Greeks coincide with ours? It needs more argument than reference to a non-existent form of perception to consolidate the objection.

2. The procedure used for establishing the peculiarities of the archaic cosmology has much in common with the method of an anthropologist who examines the world-view of an association of tribes. The differences are due to the scarcity of the evidence and to the particular circumstances of its origin (written sources; works of art; no personal contact). Let us take a closer look at this procedure!

An anthropologist trying to discover the cosmology of his chosen tribe and the way in which it is mirrored in language, in the arts, in daily life, first learns the language and the basic social habits; he inquires how they are related to other activities, including such *prima facie* unimportant activities as milking cows and cooking meals;[81] he tries to identify key ideas.[82] His attention to minutiae is not the result of a misguided urge for completeness but of the realization that what looks insignificant to one way of thinking (and perceiving) may play a most important role in another. (The differences between the paper-and-pencil operations of a

---

78. See footnote 12ff and text of the present chapter.

79. See footnote 13.

80. See Part II of Hedwig Kenner, *Das Theater und der Realismus in der Griechischen Kunst*, Vienna, 1954, especially pp. 121f.

81. Evans-Pritchard, *Social Anthropology*, New York, 1965, p. 80.

82. Ibid., p. 80.

Lorentzian and those of an Einsteinian are often minute, if discernible at all; yet they reflect a major clash of ideologies.)

Having found the key ideas the anthropologist tries to *understand* them. This he does in the same way in which he originally gained an understanding of his own language, including the language of the special profession that provides him with an income. He *internalizes* the ideas so that their connections are firmly engraved in his memory and his reactions, and can be produced at will. 'The native society has to be in the anthropologist himself and not merely in his notebooks if he is to understand it.'[83] *This process must be kept free from external interference.* For example, the researcher must not try to get a better hold on the ideas of the tribe by likening them to ideas he already knows, or finds more comprehensible or more precise. On no account must he attempt a 'logical reconstruction'. Such a procedure would tie him to the known, or to what is preferred by certain groups, and would forever prevent him from grasping the unknown world-view he is examining.

Having completed his study, the anthropologist carries within himself both the native society and his own background, and he may now start comparing the two. The comparison decides whether the native way of thinking can be reproduced in European terms (provided there is a unique set of 'European terms'), or whether it has a 'logic' of its own, not found in any Western language. In the course of the comparison the anthropologist may rephrase certain native ideas in English. This does not mean that English *as spoken independently of the comparison* already contains native ideas. It means that languages can be *bent* in many directions and that understanding does not depend on any particular set of rules.

3. The examination of key ideas passes through various stages, none of which leads to a complete clarification. Here the researcher must exercise firm control over his urge for instant clarity and logical perfection. He must never try to make a concept clearer than is suggested by the material (except as a temporary aid for further research). It is this material and not his logical intuition that determines the content of the concepts. To take an example, the Nuer, a Nilotic tribe which has been examined by Evans-Pritchard, have some interesting spatio-temporal concepts.[84] The researcher

---

83. Ibid., p. 82.

84. Evans-Pritchard, *The Nuer*, Oxford, 1940, Part III; see also the brief account in *Social Anthropology*, pp. 102ff.

who is not too familiar with Nuer thought will find the concepts 'unclear and insufficiently precise'. To improve matters he might try explicating them, using modern logical notions. That might create clear concepts, but they would no longer be Nuer concepts. If, on the other hand, he wants to get concepts which are both clear and Nuer, then he must keep his key notions vague and incomplete *until the right information comes along*, i.e. until field study turns up the missing elements which, taken by themselves, are just as unclear as the elements he has already found.

Each item of information is a building block of understanding, which means that it has to be clarified by the discovery of further blocks from the language and ideology of the tribe rather than by premature definitions. Statements such as '. . . the Nuer . . . cannot speak of time as though it was something actual, which passes, can be waited for, can be saved, and so forth. I do not think that they ever experience the same feeling of fighting against time, or of having to co-ordinate activities with an abstract passage of time, because their points of reference are mainly the activities themselves, which are generally of a leisurely character . . .'[85] are either building blocks – in this case their own content is incomplete and not fully understood – or they are preliminary attempts to anticipate the arrangement of the totality of all blocks. They must then be tested, and elucidated by the discovery of further blocks rather than by logical clarifications (a child learns the meaning of a word not by logical clarification but by realizing how it goes together with things and other words). Lack of clarity of any particular anthropological statement reflects the scarcity of the material rather than the vagueness of the logical intuitions of the anthropologist, or of his tribe.

4. Exactly the same remarks apply to any attempt to explore important modern notions such as the notion of incommensurability. Within the sciences incommensurability is closely connected with meaning. A study of incommensurability in the sciences will therefore produce statements that contain meaning-terms – but these terms will be only incompletely understood, just as the term 'time' is incompletely understood in the quotation of the preceding paragraph. Thus the remark that such statements should be made only *after* production of a clear theory of

---

85. *The Nuer*, p. 103.

meaning[86] is as sensible as the remark that statements about Nuer time, which are the material that *leads to* an understanding of Nuer time, should be written down only after such an understanding has been achieved.

5. Logicians are liable to object. They point out that an examination of meanings and of the relation between terms is the task of *logic*, not of anthropology. Now by 'logic' one may mean at least two different things. 'Logic' may mean the study of, or results of the study of, the structures inherent in a certain type of discourse. And it may mean a particular logical system, or set of systems.

A study of the first kind belongs to anthropology. For in order to see, for example, whether $AB \vee AB \equiv A$ is part of the 'logic of quantum theory' we shall have to study quantum theory. And as quantum theory is not a divine emanation but a human product, we shall have to study it in the form in which human products usually are available, that is, we shall have to study historical records – textbooks, original papers, records of meetings and private conversations, letters, and the like. (In the case of quantum theory our position is improved by the fact that the tribe of quantum theoreticians has not yet died out. Thus we can supplement historical study with anthropological field work such as the work of Kuhn and his collaborators.[87])

It is to be admitted that these records do not, by themselves, produce a *unique* solution to our problems. But who has ever assumed that they do? Historical records do not produce a unique solution for historical problems either, and yet nobody suggests that they be neglected. There is no doubt that the records are *necessary* for a logical study in the sense examined now. The question is how they should be *used*.

We want to discover the structure of the field of discourse, of which the records give an incomplete account. We want to learn about it without changing it in any way. In our example we are not interested in whether a *perfected* quantum mechanics of the future employs $AB \vee AB \equiv A$ or

---

86. Achinstein, *Minnesota Studies in the Philosophy of Science*, Vol. 4, Minneapolis, 1970, p. 224, says that 'Feyerabend owe[s] us a theory of meaning' and Hempel is prepared to accept incommensurability only *after* the notion of meaning involved in it has been made clear, op. cit., p. 156.

87. Report in T.S. Kuhn, J.L. Heilbron, P. Forman and L. Allen, *Sources for the History of Quantum Physics*, American Philosophical Society, Philadelphia, 1967. The material assembled under the programme described in this report can be consulted at various universities, the University of California in Berkeley among them.

whether an *invention* of our own, whether a little bit of 'reconstruction' which changes the theory so that it conforms to some preconceived principles of modern logic and readily provides the answer employs that principle. We want to know whether quantum theory *as actually practised by physicists* employs the principle. For it is the work of the physicists and not the work of the reconstructionists we want to examine. And this work may well be full of contradictions and lacunae. Its 'logic' (in the sense in which I am now using the term) may well be 'illogical' when judged from the point of view of a particular system of formal logic.

Putting the question in this way we realize that it may not admit of any answer. There may not exist a single theory, one 'quantum theory', that is used in the same way by all physicists. The difference between Bohr, Dirac, Feynman and von Neumann suggests that this is more than a distant possibility. To test the possibility, i.e. to either eliminate it or to give it shape, we must examine concrete cases. Such an examination of concrete cases may then lead to the result that quantum theoreticians differ from each other as widely as do Catholics and the various types of Protestants: they may use the same texts (though even that is doubtful – just compare Dirac with von Neumann), but they sure are doing different things with them.

The need for anthropological case studies in a field that initially seemed to be dominated by a single myth, always the same, always used in the same manner, indicates that our common knowledge of science may be severely defective. It may be entirely mistaken (some mistakes have been hinted at in the preceding chapters). In these circumstances, the only safe way is to confess ignorance, to abandon reconstructions, and to start studying science from scratch. We must approach science like an anthropologist approaches the mental contortions of the medicine-men of a newly discovered association of tribes. And we must be prepared for the discovery that these contortions *are* wildly illogical (when judged from the point of view of a particular system of formal logic) and *have to be* wildly illogical in order to function as they do.

6. Only a few philosophers of science interpret 'logic' in this sense, however. Only few philosophers are prepared to concede that the basic structures that underlie some newly discovered idiom might differ radically from the basic structures of the more familiar systems of formal logic, and absolutely nobody is ready to admit that this might be true of science as well. Most of the time the 'logic' (in the sense discussed so far) of a particular language, or of a theory, is immediately identified with the features of

a particular logical system without considering the need for an inquiry concerning the adequacy of such an identification. Professor Giedymin, for example,[88] means by 'logic' a favourite system of his which is fairly comprehensive, but by no means all-embracing. (For example, it does not contain, nor could it be used to formulate, Hegel's ideas. And there have been mathematicians who have doubted that it can be used for expressing informal mathematics.) A logical study of science as Giedymin and his fellow logicians understand it is a study of sets of formulae of this system, of their structure, the properties of their ultimate constituents (intension, extension, etc.), of their consequences and of possible models. If this study does not repeat the features an anthropologist has found in, say, science then this either shows that science has some faults, or that the anthropologist does not know any logic. It does not make the slightest difference to the logician in this second sense that his formulae *do not look* like scientific statements, that they *are not used* like scientific statements and that science could not possibly grow in the simple ways his brain is capable of understanding (and therefore regards as the only permissible ways). He either does not notice the discrepancy or he regards it as being due to imperfections that cannot enter a satisfactory account. Not once does it occur to him that the 'imperfections' might have a positive *function*, and that scientific progress might be impossible once they are removed. For him science *is* axiomatics plus model theory plus correspondence rules plus observation language.

Such a procedure assumes (without noticing that there is an assumption involved) that an anthropological study which familiarizes us with the overt and the hidden classifications of science has been completed, and that it has decided in favour of the axiomatic (etc., etc.) approach. No such study has ever been carried out. And the bits and pieces of field work available today, mainly as the result of the work of Hanson, Kuhn, Lakatos and the numerous historians who remained untouched by positivistic prejudices, show that the logician's approach removes not just some inessential embroideries of science, but those very features which make scientific progress and thereby science possible.

7. The discussions of meaning I have alluded to are another illustration of the deficiencies of the logician's approach. For Giedymin, this term and its derivatives, such as the term 'incommensurability', are 'unclear and

---

88. *British Journal for the Philosophy of Science*, August 1970, pp. 257ff and February 1971, pp. 39ff.

insufficiently precise'. I agree. Giedymin wants to make the terms clearer, he wants to understand them better. Again agreement. He tries to obtain the clarity he feels is lacking by explication in terms of a particular kind of formal logic and of the double language model, restricting the discussion to 'intension' and 'extension' as explained in the chosen logic. It is here that the disagreement starts. For the question is not how 'meaning' and 'incommensurability' occur within a particular logical system. The question is what role they play in (actual, i.e. non-reconstructed) science. Clarification must come from a more detailed study of this role, and lacunae must be filled with the results of such study. And as the filling takes time the key terms will be 'unclear and insufficiently precise' for years and perhaps decades. (See also items 3 and 4 above.)

8. Logicians and philosophers of science do not see the situation in this way. Being both unwilling and unable to carry out an informal discussion, they demand that the main terms of the discussion be 'clarified'. And to 'clarify' the terms of a discussion does not mean to study the *additional* and as yet unknown properties of the domain in question which one needs to make them fully understood, it means to fill them with *existing* notions from the entirely different domain of logic and common sense, preferably observational ideas, until they sound common themselves, and to take care that the process of filling obeys the accepted laws of logic. The discussion is permitted to proceed only *after* its initial steps have been modified in this manner. So the course of an investigation is deflected into the narrow channels of things already understood and the possibility of fundamental conceptual discovery (or of fundamental conceptual change) is considerably reduced. Fundamental conceptual change, on the other hand, presupposes new world-views and new languages capable of expressing them. Now, building a new world-view, and a corresponding new language, is a process that takes time, in science as well as in meta-science. The terms of the new language become clear only when the process is fairly advanced, so that each single word is the centre of numerous lines connecting it with other words, sentences, bits of reasoning, gestures which sound absurd at first but which become perfectly reasonable once the connections are made. Arguments, theories, terms, points of view and debates can therefore be clarified in at least two different ways: (a) in the manner already described, which leads back to the familiar ideas and treats the new as a special case of things already understood, and (b) by incorporation into a language of the future, which means *that one must learn to argue with unexplained terms*

and to use sentences for which no clear rules of usage are as yet available. Just as a child who starts using words without yet understanding them, who adds more and more uncomprehended linguistic fragments to his playful activity, discovers the sense-giving principle only *after* he has been active in this way for a long time – the activity being a necessary presupposition of the final blossoming forth of sense – in the very same way the inventor of a new world-view (and the philosopher of science who tries to understand his procedure) must be able to talk nonsense until the amount of nonsense created by him and his friends is big enough to give sense to all its parts. There is again no better account of this process than the description which John Stuart Mill has left us of the vicissitudes of his education. Referring to the explanations which his father gave him on logical matters, he wrote: 'The explanations did not make the matter at all clear to me at the time; but they were not therefore useless; they remained as a nucleus for my observations and reflections to crystallise upon; the import of his general remarks being interpreted to me, by the particular instances which came under my notice *afterwards*.'[89] Building a new language (for understanding the world, or knowledge) is a process of exactly the same kind *except* that the initial 'nuclei' are not given, but must be invented. We see here how essential it is to learn talking in riddles, and how disastrous an effect the drive for instant clarity must have on our understanding. (In addition, such a drive betrays a rather narrow and barbaric mentality: 'To use words and phrases in an easy going way without scrutinizing them too curiously is not, in general, a mark of ill breeding; on the contrary, there is something low bred in being too precise. . . .'[90])

All these remarks are rather trivial and can be illustrated by obvious examples. Classical logic arrived on the scene only when there was sufficient argumentative material (in mathematics, rhetoric, politics) to serve as a starting point and as a testing ground. Arithmetic developed without any clear understanding of the concept of number; such

---

89. There is much more randomness in this process than a rationalist would ever permit, or suspect, or even notice. See von Kleist, 'Über die allmähliche Verfertigung der Gedanken beim Reden', in *Meisterwerke Deutscher Literaturkritik*, ed. Hans Meyer, Stuttgart, 1962, pp. 741–7. Hegel had an inkling of the situation. See K. Loewith and J. Riedel (eds), *Hegel, Studienausgabe I*, Frankfurt, 1968, p. 54. For Mill See Chapter 11, footnote 13.

90. Plato, *Theaitetos*, 184c. See also I. Düring, *Aristoteles*, Heidelberg, 1966, p. 379, criticizing Aristotle's demand for instant precision.

understanding arose only when there existed a sufficient amount of arithmetical 'facts' to give it substance. In the same way a proper theory of meaning (and of incommensurability) can arise only after a sufficient number of 'facts' has been assembled to make such a theory more than an exercise in concept-pushing. This is the reason for the examples in the present section.

9. There is still another dogma to be considered before returning to the main narration. It is the dogma that all subjects, however assembled, quite automatically obey the laws of logic, or ought to obey the laws of logic. If this is so, then anthropological field work would seem to be superfluous. 'What is true in logic is true in psychology . . . in scientific method, and in the history of science,' writes Popper.[91]

This dogmatic assertion is neither clear nor is it (in one of its main interpretations) true. To start with, assume that the expressions 'psychology', 'history of science', 'anthropology' refer to certain domains of facts and regularities (of nature, of perception, of the human mind, of society). Then the assertion is not *clear* as there is not a single subject – LOGIC – that underlies all these domains. There is Hegel, there is Brouwer, there are the many logical systems considered by modern constructivists. They offer not just different interpretations of one and the same bulk of logical 'facts', but different 'facts' altogether. And the assertion is not *true* as there exist legitimate scientific statements which violate simple logical rules. For example, there are statements which play an important role in established scientific disciplines and which are observationally adequate only if they are self-contradictory: fixate a moving pattern that has just come to a standstill, and you will see it move in the opposite direction, but without changing its position. The only phenomenologically adequate description is 'it moves, in space, but it does not change place' – and this description is self-contradictory.[92]

---

91. *Objective Knowledge*, Oxford, 1972, p. 6. Anticipated e.g. by Comte, *Course*, 52° Leçon and, of course, Aristotle.

92. It has been objected (Ayer, G.E.L. Owen) that we are dealing with appearances, not with actual events, and that the correct description is 'it appears to move. . . .' But the difficulty remains. For if we introduce the 'appear', we must put it at the beginning of the sentence, which will read 'it appears that it moves and does not change place'. And as appearances belong to the domain of phenomenological psychology we have made our point, viz. that this domain contains self-inconsistent elements.

There are examples from geometry:[93] thus the enclosed figure (which need not appear in the same way to every person) is seen as an isosceles triangle whose base is not halved by the perpendicular. And there are examples with a = b & b = c & a ≥ c as the only phenomenologically adequate description.[94] Moreover, there is not a single science, or other form of life that is useful, progressive as well as in agreement with logical demands. Every science contains theories which are inconsistent both with facts and with other theories and which reveal contradictions when analysed in detail. Only a dogmatic belief in the principles of an allegedly uniform discipline 'Logic' will make us disregard this situation. And the objection that logical principles and principles of, say, arithmetic differ from empirical principles by not being accessible to the method of conjecture and refutations (or, for that matter, any other 'empirical' method) has been defused by more recent research in this field.[95]

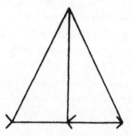

Secondly, let us assume that the expressions 'psychology', 'anthropology', 'history of science', 'physics' do not refer to facts and laws, but to certain *methods* of assembling facts including certain ways of connecting observation with theory and hypothesis. That is, let us consider the *activity* 'science' and its various subdivisions. Then we may lay down *ideal demands* of knowledge and knowledge-acquisition, and we may try to construct a (social) machinery that obeys these demands.

93. E. Rubin, 'Visual Figures Apparently Incompatible with Geometry', *Acta Psychologica*, VII, 1950, pp. 365ff. Cf. also the drawings on p. 166–7.

94. E. Tranekjaer-Rasmussen, 'Perspectoid Distances', *Acta Psychologica*, XI, 1955, p. 297.

95. Mainly by the work of Imre Lakatos, 'Proofs and Refutation', *British Journal for the Philosophy of Science*, 1962/63.

Almost all epistemologists and philosophers of science proceed in this way. Occasionally they succeed in finding a machinery that might work in certain ideal conditions, but they never inquire, or even find it worth inquiring, whether the conditions are satisfied in this real world of ours. Such an inquiry, on the other hand, will have to explore the way in which scientists *actually* deal with their surroundings, it will have to examine the actual shape of their product, viz. 'knowledge', and the way in which this product changes as a result of decisions and actions in complex social and material conditions. In a word, such an inquiry will have to be anthropological.

There is no way of predicting what an anthropological inquiry will bring to light. In the preceding chapters, which are rough sketches of an anthropological study of particular episodes, it has emerged that science is full of lacunae and contradictions, that ignorance, pigheadedness, reliance on prejudice, lying, far from impeding the forward march of knowledge may actually aid it, and that the traditional virtues of precision, consistency, 'honesty', respect for facts, maximum knowledge under given circumstances, if practised with determination, may bring it to a standstill. It has also emerged that logical principles not only play a much smaller role in the (argumentative and non-argumentative) moves that advance science, but that the attempt to enforce them would seriously impede science. (One cannot say that von Neumann has advanced the quantum theory. But he certainly made the discussion of its basis more long-winded and cumbersome.[96])

Now a scientist engaged in a certain piece of research has not yet completed all the steps that lead to definite results. His future is still open. Will he follow the barren and illiterate logician who preaches to him about the virtues of clarity, consistency, experimental support (or experimental falsification), tightness of argument, 'honesty', and so on, or will he imitate his predecessors in his own field who advanced by breaking most of the rules logicians want to lay on him? Will he rely on abstract injunctions or on the results of a study of concrete episodes? I think the answer is clear, and with it the relevance of anthropological field work not just for the anthropologists but also for the members of the societies he examines. I now continue my narration and proceed

---

96. Besides, the imprecisions which he removes from the formalism now reappear in the relation between theory and fact. Here the correspondence principle still reigns supreme. See footnote 25 of Chapter 5.

to describing the transition from the paratactic universe of the archaic Greeks to the substance-appearance universe of their followers.

The archaic cosmology (which from now on I shall call cosmology A) contains things, events, their parts; it does not contain appearances.[97] Complete knowledge of an object is complete enumeration of its parts and peculiarities. Humans cannot have complete knowledge. There are too many things, too many events, too many situations (*Iliad*, 2.488), and they can be close to only a few of them (*Iliad*, 2.485). But although humans cannot have complete knowledge, they can have a sizeable amount of it. The wider their experience, the greater the number of adventures, of things seen, heard, read, the greater their knowledge.[98]

The new cosmology (cosmology B) that arises in the 7th to 5th centuries BC distinguishes between much-knowing, πολυμαθίη, and true knowledge,[99] and it warns against trusting 'custom born of manifold experience', εθος πολύπειεον.[100] Such a distinction and such a warning make sense only in a world whose structure differs from the structure of A. In one version which played a large role in the development of Western civilization and which underlies such problems as the problem of the existence of theoretical entities and the problem of alienation, the new events form what one might call a *True World*, while the events of everyday life are now *appearances* that are but its dim and misleading reflection.[101] The True World is simple and coherent, and it can be

97. Snell, *Ausdrücke*, p. 28 (referring to Homer), speaks of a 'knowledge that proceeds from appearances and draws their multitude together in a unit which is then posited as their true essence'. This may apply to the Presocratics, it does not apply to Homer. In the case of Homer 'the world is comprehended as the sum of things, visible in space, and not as reason acting intensively' (ibid., p. 67, discussing Empedokles; see also the lines following the quotation for a further elaboration of the theme).

98. Snell, *Die alten Griechen und Wir*, p. 48.

99. See Heraclitus, fr. 40 (Diels–Kranz).

100. Parmenides, fr. 7, 3. 'Here for the first time sense and reason are contrasted'; W.K. Guthrie, *A History of Greek Philosophy*, Vol. II, Cambridge, 1965, p. 25.

101. This distinction is characteristic of certain mythological views as well. Homer thus differs both from the preceeding mythologies and from the succeeding philosophies. His point of view is of great originality. In the 20th century J.L. Austin has developed similar ideas. And he has criticized the development from Thales via Plato to the present essentialism. See the first chapter of *Sense and Sensibilia*. Chapter 3 of *Farewell to Reason* contains details.

described in a uniform way. So can every act by which its elements are comprehended: a few abstract notions replace the numerous concepts that were used in cosmology A for describing how humans might be 'inserted' into their surroundings and for expressing the equally numerous types of information thus gained. From now on there is only one important type of information, and that is: *knowledge*.

The conceptual totalitarianism that arises as a result of the slow arrival of world B has interesting consequences, not all of them desirable. Situations which made sense when tied to a particular type of cognition now become isolated, unreasonable, apparently inconsistent with other situations: we have a 'chaos of appearances'. The 'chaos' is a direct consequence of the simplification of language that accompanies the belief in a True World.[102] Moreover, all the manifold abilities of the observers are now directed towards this True World, they are adapted to a *uniform* aim, shaped for *one particular* purpose; they become more similar to each other, which means that humans become impoverished together with their language. They become impoverished at precisely the moment they discover an autonomous 'I' and proceed to what some have been pleased to call a 'more advanced notion of God' (allegedly found in Xenophanes), which is a notion of God lacking the rich variety of typically human features.[103] 'Mental' events which before were treated in analogy with events of the body and which *were experienced accordingly*[104] become more 'subjective', they become modifications, actions, revelations of a spontaneous soul: the distinction between appearance (first impression, mere opinion) and reality (true knowledge) spreads everywhere. Even the task of the artist now consists in arranging his shapes in such a manner that the underlying essence can be grasped with ease. In painting this leads to the development of what one can only call systematic methods for deceiving the eye: the archaic artist treats the surface on which he paints as a writer might treat a piece of papyrus; it *is* a real surface, it is supposed to be *seen* as a real surface

---

102. Snell, *Ausdrücke*, pp. 80f; von Fritz, *Philosophie und sprachlicher Ausdruck bei Demokrit, Plato und Aristoteles*, Leipzig–Paris–London, 1938, p. 11.

103. '. . . in becoming the embodiment of cosmic justice Zeus lost his humanity. Hence Olympianism in its moralized form tended to become a religion of fear . . .', Dodds, *Greeks*, p. 35. For Xenophanes see Chapter 2 of *Farewell to Reason*.

104. Snell, *Discovery*, p. 69.

(though attention is not always directed to it) and the marks he draws on it are comparable to the lines of a blueprint or the letters of a word. They are symbols that *inform* the reader of the *structure of the object*, of its parts, of the way in which the parts are related to each other. The simple drawing below, for example, may represent three paths meeting at a point. The artist using perspective, on the other hand, regards the surface and the marks he puts on it as *stimuli* that trigger the *illusion* of an arrangement of three-dimensional objects. The illusion occurs because the human mind is capable of producing illusory experiences when properly stimulated. The drawing is now seen either as the corner of a cube that extends towards the viewer, or as the corner of a cube that points away from him (and is seen from below), or else as a plane floating above the surface of the paper carrying a two-dimensional drawing of three paths meeting.

Combining this new way of seeing with the new concept of knowledge that has just been described, we obtain new entities, viz. physical objects as they are understood by most contemporary philosophers. To explain, let me again take the case of the oar.

In the archaic view 'the oar' is a complex consisting of parts some of which are objects, some situations, some events. It is possible to say 'the straight oar is broken' (*not* 'appears to be broken') just as it is possible to say 'swift-footed Achilles is walking slowly', for the elements are not set against each other. They are part of a paratactic aggregate. Just as a traveller explores all parts of a strange country and describes them in a 'periegesis' that enumerates its peculiarities, one by one, in the same way the student of simple objects such as oars, boats, horses, people inserts himself into the 'major oar-situations', apprehends them in the appropriate way, and reports them in a list of properties, events, relations. And just as a detailed periegesis exhausts what can be said about a country, in the same way a detailed list exhausts what can be said about an object.[105] 'Broken in water' belongs to the oar as does 'straight to the hand'; it is 'equally real'. In cosmology B, however, 'broken in water' is a 'semblance' that *contradicts* what is suggested by the 'semblance' of straightness and thus shows the basic untrustworthiness of all semblances.[106] The concept of an object has changed from the concept of an aggregate of equi-important perceptible parts to the concept of an imperceptible essence underlying a multitude of deceptive phenomena. (We may guess that the appearance of an object has changed in a similar way, that objects now look less 'flat' than before.)

Considering these changes and peculiarities, it is plausible to assume that the comparison of A and B *as interpreted by the participants* (rather than as 'reconstructed' by logically well-trained but otherwise illiterate outsiders) will raise various problems. In the remainder of this chapter only some aspects of some of these problems will be discussed. Thus I shall barely mention the psychological changes that accompany the transition from A to B and which are not just a matter of conjecture, but can be established by independent research. Here is rich material for the detailed study of the role of frameworks (mental sets, languages, modes of representation) and the limits of rationalism.

To start with, cosmos A and cosmos B are built from different *elements*.

The elements of A are relatively independent parts of objects which enter into external relations. They participate in aggregates without changing their intrinsic properties. The 'nature' of a particular aggregate is determined by its parts and by the way in which the parts are related to each other. *Enumerate the parts in the proper order, and you have the object.* This applies to physical aggregates, to humans (minds and bodies), to animals, but it also applies to social aggregates such as the honour of a warrior.

The elements of B fall into two classes: essences (objects) and appearances (of objects – what follows is true only of some rather streamlined versions of B). Objects (events, etc.) may again combine. They may form harmonious

---

105. The idea that knowledge consists in *lists* reaches back far into the Sumerian past. See von Soden, *Leistung und Grenzen Sumerisch-Babylonischer Wissenchaft*, new edn, Darmstadt, 1965. The difference between Babylonian and Greek mathematics and astronomy lies precisely in this. The one develops methods for the presentation of what we today call 'phenomena' and which were interesting and relevant events in the sky, while the other tries to develop astronomy, 'while leaving the heavens alone' (Plato, *Rep.*, 530bf; *Lgg.*, 818a).

106. Xenophanes, fr. 34.

totalities where each part gives meaning to the whole and receives meaning from it (an extreme case is Parmenides, where isolated parts are not only unrecognizable but altogether unthinkable). Aspects properly combined do not produce *objects*, but psychological conditions for the apprehension of *phantoms* which are but other aspects, and particularly misleading ones at that (they look so convincing). *No enumeration of aspects is identical with the object* (problem of induction).

The transition from A to B thus introduces new entities and new relations between entities (this is seen very clearly in painting and statuary). It also changes the concept and the self-experience of humans. An archaic individual is an assemblage of limbs, connections, trunk, neck, head,[107] (s)he is a puppet set in motion by outside forces such as enemies, social circumstances, feelings (which are described and perceived as objective agencies – see above):[108] 'Man is an open target of a great many forces which impinge on him, and penetrate his very core.'[109] He is an exchange station of material and spiritual, but always objective, causes. And this is not just a 'theoretical' idea, it is a social fact. Man is not only *described* in this way, he is *pictured* in this way, and he *feels* himself to be constituted in this manner. He does not possess a central agency of action, a spontaneous 'I' that produces *its own* ideas, feelings, intentions, and differs from behaviour, social situations, 'mental' events of type A. Such an I is neither mentioned nor is it noticed. It is nowhere to be found within A. But it plays a very decisive role within B. Indeed, it is not implausible to assume that some outstanding peculiarities of B such as aspects, semblances, ambiguity of feeling[110] enter the stage as a result of a *sizeable increase of self-consciousness*.

Now one might be inclined to explain the transition as follows: archaic man has a limited cosmology; he discovered some things, he missed others. His universe lacks important objects, his language lacks important concepts, his perception lacks important structures. Add the missing elements to cosmos A, the missing terms to language A, the

---

107. 'To be precise, Homer does not even have any words for the arms and the legs; he speaks of hands, lower arms, upper arms, feet, calves, and thighs. Nor is there a comprehensive term for the trunk.' Snell, *Discovery*, Chapter 1, footnote 7.

108. 'Emotions do not spring spontaneously from man, but are bestowed on him by the gods,' ibid., p. 52. See also the account earlier in the present chapter.

109. Ibid., p. 20.

110. See Sappho's 'bitter-sweet Eros', ibid, p. 60.

missing structures to the perceptual world of A, and you obtain cosmos B, language B, perception B.

Some time ago I called the theory underlying such an explanation the 'hole theory' or the 'Swiss cheese theory' of language (and other means of representation). According to the hole theory every cosmology (every language, every mode of perception) has sizeable lacunae which can be filled, *leaving everything else unchanged*. The hole theory is beset by numerous difficulties. In the present case there is the difficulty that cosmos B does not contain a single element of cosmos A. Neither common-sense terms, nor philosophical theories; neither painting and statuary, nor artistic conceptions; neither religion, nor theological speculation contain a single element of A once the transition to B has been completed. *This is a historical fact.*[111] Is this fact an accident, or has A some structural properties that prevent the co-existence of A-situations and B-situations? Let us see!

I have already mentioned an example that might give us an inkling of a reason as to why B does not have room for A-facts: the drawing below may be the intersection of three paths as presented in accordance with the principles of A-pictures (which are visual lists). Perspective having been introduced (either as an objective method or as a mental set), it can no longer be seen in this manner. Instead of lines on paper we have the illusion of depth and a three-dimensional panorama, though of a rather simple kind. There is no way of incorporating the A-picture into the B-picture except as part of this illusion. But an illusion of a visual list is not a visual list.

---

111. The fact is not easy to establish. Many presentations of A, including some very detailed and sophisticated ones, are infected by B-concepts. An example is quoted in footnote 97 to the present chapter. Here as elsewhere only the anthropological method can lead to knowledge that is more than a reflection of wishful thinking. A similar situation in the course of individual development is described in the text to footnote. 12 of the present chapter.

The situation becomes more transparent when we turn to concepts. I have said above that the 'nature' of an object (= aggregate) in A is determined by the elements of the aggregate and the relation between the elements. One should add that this determination is 'closed' in the sense that the elements and their relations *constitute* the object; when they are given, then the object is given as well. For example, the 'elements' described by Odysseus in his speech in *Iliad*, 9.225ff *constitute* honour, grace, respect. A-concepts are thus very similar to notions such as 'checkmate': given a certain arrangement of pieces on the board, there is no way of 'discovering' that the game can still be continued. Such a 'discovery' would not fill a gap, it would not add to our knowledge of possible chess positions, it would put an end to the game. And so would the 'discovery' of 'real meanings' behind other moves and other constellations.

Precisely the same remarks apply to the 'discovery' of an individual I that is different from faces, behaviour, objective 'mental states' of the type that occur in A, to the 'discovery' of a substance behind 'appearances' (formerly elements of A), or to the 'discovery' that honour may be lacking despite the presence of all its outer manifestations. A statement such as Heraclitus' 'you could not find the limits of the soul though you are travelling every way, so deep is its *logos*' (Diels, B 45) does not just *add* to cosmos A, it *undercuts* the principles which are needed in the construction of A-type 'mental states' while Heraclitus' rejection of πολυμαθίη and Parmenides' rejection of an εθοϑ πολπειον undercuts rules that govern the construction of *every single fact* of A. An entire world-view, an entire universe of thought, speech, perception is dissolved.

It is interesting to see how this process of dissolving manifests itself in particular cases. In his long speech in *Iliad*, 9.308ff, Achilles wants to say that honour may be absent even though all its outer manifestations are present. The terms of the language he uses are so intimately tied to definite social situations that he 'has no language to express his disillusionment. Yet he expresses it, and in a remarkable way. He does it by misusing the language he disposes of. He asks questions that cannot be answered and makes demands that cannot be met.'[112] He acts in a most 'irrational' way.

---

112. A. Parry, 'The Language of Achilles', *Trans. & Proc. Amer. Phil. Assoc*, 87, 1956, p. 6. Cf. the discussion of the case in *Farewell to Reason*, Chapter 10.

The same irrationality is found in the writings of all other early authors. Compared with A the Presocratics speak strangely indeed. So do the lyrical poets who explore the new possibilities of selfhood they have 'discovered'. Freed from the fetters of a well-constructed and unambiguous mode of expression and thinking, the elements of A lose their familiar function and start floating around aimlessly – the 'chaos of sensations' arises. Freed from firm and unambiguous social situations feelings become fleeting, ambivalent, contradictory: 'I love, and I love not; I rave, and I do not rave,' writes Anakreon.[113] Freed from the rules of late geometric painting the artists produce strange mixtures of perspective and blueprint.[114] Separated from well-determined psychological sets and freed of their realistic import, concepts may now be used 'hypothetically' without any odium of lying, and the arts may begin exploring possible worlds in an imaginative way.[115] This is the same 'step back' which was earlier seen to be a necessary presupposition of change and, possibly, progress[116] – only it now does not just discard observations, it discards some important standards of rationality as well. Seen from A (and also from the point of view of some later ideologies) all these thinkers, poets, artists, are raving maniacs.

---

113. Diehl, *Anthologia Lyrica*, fr. 79.

114. Pfuhl, op. cit.; see also J. White, *Perspective in Ancient Drawing and Painting*, London, 1965.

115. Plutarch reports the following story in his *Life of Solon*: 'When the company of Thespis began to exhibit tragedy, and its novelty was attracting the populace but had not yet got as far as public competitions, Solon, being fond of listening and learning and being rather given in his old age to leisure and amusement, and indeed to drinking parties and music, went to see Thespis act in his own play, as was the practice in ancient times. Solon approached him after the performance and asked him if he was not ashamed to tell so many lies to so many people. When Thespis said there was nothing dreadful in representing such works and actions in fun, Solon struck the ground violently with his walking stick: "If we applaud these things in fun," he said, "we shall soon find ourselves honouring them in earnest". The story seems historically impossible yet elucidates a widespread attitude (for this attitude see Chapter 8 of John Forsdyke *Greece before Homer*, New York, 1964). Solon himself seems to have been somewhat less impressed by traditional forms of thought and he may have been one of the first dramatic actors (of the political variety): G. Else, *The Origin and Early Form of Tragedy*, Cambridge, 1965, pp. 40ff. The opposite attitude, which reveals the secure and already somewhat conceited citizen of B, is expressed by Simonides who answered the question why the Thessalians were not deceived by him by saying 'Because they are too stupid'. Plutarch, *De aud. poet.*, 15D.

116. Chapter 11, text to footnote 5.

Remember the circumstances which are responsible for this situation. We have a point of view (theory, framework, cosmos, mode of representation) whose elements (concepts, 'facts', pictures) are built up in accordance with certain principles of construction. The principles involve something like a 'closure': there are things that cannot be said, or 'discovered', without violating the principles (which does *not* mean contradicting them). Say the things, make the discovery, and the principles are suspended. Now take those constructive principles that underlie every element of the cosmos (of the theory), every fact (every concept). Let us call such principles *universal principles* of the theory in question. Suspending universal principles means suspending all facts and all concepts. Finally, let us call a discovery, or a statement, or an attitude *incommensurable* with the cosmos (the theory, the framework) if it suspends some of its universal principles. Heraclitus B 45 is incommensurable with the psychological part of A: it suspends the rules that are needed for constituting individuals and puts an end to all A-facts about individuals (phenomena corresponding to such facts may of course persist for a considerable time as not all conceptual changes lead to changes in perception, and as there exist conceptual changes that never leave a trace in the appearances; however, such phenomena can no longer be *described* in the customary way and cannot therefore count as observations of the customary 'objective facts').

Note the tentative and vague nature of this explanation of 'incommensurable' and the absence of logical terminology. The reason for the vagueness has already been explained (items 3 and 4 above). The absence of logic is due to the fact that we deal with phenomena outside of its domain. My purpose is to find terminology for describing certain complex historical-anthropological phenomena which are only imperfectly understood rather than defining properties of logical systems that are specified in detail. Terms, such as 'universal principles' and 'suspend', are supposed to summarize anthropological information much in the same way in which Evans-Pritchard's account of Nuer time (text to footnote 85) summarizes the anthropological information at his disposal (see also the brief discussion in item 3 above). The vagueness of the explanation reflects the incompleteness and complexity of the material and invites articulation by further research. The explanation has to have *some* content – otherwise it would be useless. But it must not have *too much* content, or else we have to revise it every second line.

Note, also, that by a 'principle' I do not simply mean a *statement* such as 'concepts apply when a finite number of conditions is satisfied', or 'knowledge is enumeration of discrete elements which form paratactic aggregates' but the *grammatical habit* corresponding to the statement. The two statements just quoted describe the habit of regarding an object as given when the list of its parts has been fully presented. This habit is suspended (though not contradicted) by the *conjecture* that even the most complete list does not exhaust an object; it is *also* suspended (but again not contradicted) by any unceasing search for new aspects and new properties. (It is therefore not feasible to define 'incommensurability' by reference to statements.[117]) If the habit is suspended, then A-objects are suspended with it: one cannot examine A-objects by a method of conjectures and refutations that knows no end.

How is the 'irrationality' of the transition period overcome? It is overcome in the usual way (see item 8 above), i.e. by the determined production of nonsense until the material produced is rich enough to permit the rebels to reveal, and everyone else to recognize, new universal principles. (Such revealing need not consist in writing the principles down in the form of clear and precise statements.) Madness turns into sanity provided it is sufficiently rich and sufficiently regular to function as the basis of a new world-view. And when *that* happens, then we have a new problem: How can the old view be compared with the new view?

From what has been said it is obvious that we cannot compare the *contents* of A and B. A-facts and B-facts cannot be put side by side, not even in memory: presenting B-facts means suspending principles assumed in the construction of A-facts. All we can do is draw B-pictures of A-facts in B, or introduce B-statements of A-facts into B. We cannot use A-statements of A-facts in B. Nor is it possible to *translate* language A into language B. This does not mean that we cannot *discuss* the two views – but the discussion will lead to sizeable changes of both views (and of the languages in which they are expressed).

Now it seems to me that the relation between, say, classical mechanics (interpreted realistically) and quantum mechanics (interpreted in accordance with the views of Niels Bohr), or between Newtonian

---

117. This takes care of a criticism in footnote 63 of Shapere's article in *Mind and Cosmos*, Pittsburgh, 1966. The classifications achieved by the principles are 'covert' in the sense of Whorf: see above, footnote 4 and text down to footnote 9.

mechanics (interpreted realistically) and the general theory of relativity (also interpreted realistically) is in many respects similar to the relation between cosmology A and cosmology B. Thus every fact of Newton's mechanics presumes that shapes, masses, periods are changed only by physical interactions, and this presumption is suspended by the theory of relativity. Similarly the quantum theory constitutes facts in accordance with the uncertainty relations which are suspended by the classical approach.

At this point it is important to interpret the situation in a sensible manner, or else scientific (cultural) change becomes an inexplicable miracle. The idea that comprehensive ways of thinking, acting, perceiving such as cosmology A (and, in a much more narrow domain, classical physics) and cosmology B (relativity or quantum mechanics) are closed frameworks with fixed rules creates an unbridgeable gulf between situations which, though different in surprising ways, are yet connected by arguments, allusions, borrowings, analogies, general principles of the kind explained in the text above. Logicians who confine the term 'argument' to chains of reasoning involving stable and precise concepts and who reconstruct theories and world-views using equally precise and unambiguous terms are forced to call such connections 'irrational', while their opponents can report the 'discovery' that science, that alleged stronghold of reason, often violates reason in a decisive way. Both are talking about chimaeras, not about science and culture as they really are. Things change when we use scientific practice or cultural reality and not logic as our informants, in other words, when we engage in sociological research, not in reconstruction. We then discover that scientific concepts (and concepts, shapes, percepts, styles in general) are ambiguous in the sense that decisive events can affect their appearance, their perceived implications and, with them, the 'logic' they obey. Achilles (see the text to footnote 112 above) 'misuses' the language he has at his disposal by asserting a difference between 'real' honour and its social manifestations. Asserting differences is not in conflict with view A; for example, there is a great difference between the knowledge, the power, the actions of the gods on the one side and the knowledge, the power and the actions of humans on the other. Assuming that honour is in the hands of gods who don't give a damn about the aspirations of humans devalues the social manifestations of honour, makes them secondary. The assumption fits well into the general outlines of view A, but Achilles is the first to make it.

Why? Because his anger, his suffering makes him see connections which, because of a widespread optimism, are not part of the general views about honour and do not contribute to its 'definition'. He seems to violate basic social rules but viewed with the anxiety caused by Agamemnon's actions such rules, give way to a different idea that is regarded as being implicit in the existing material but as not having surfaced so far. Generalizing, we can say that concepts have potentialities over and above the usages that seem to define them; it is this feature that makes them capable of connecting entirely different conceptual systems. More about this in my (I promise!) last book, *The Conquest of Abundance*.

Appendix 2

Whorf speaks of 'Ideas', not of 'events' or of 'facts', and it is not always clear whether he would approve of my extension of his views. On the one hand he says that 'time, velocity, and matter are not essential to the construction of a consistent picture of the universe,[1] and he asserts that 'we cut up nature, organize it into concepts, and ascribe significances as we do, largely because we are partial to an agreement to organize it in this way',[2] which would seem to imply that widely different languages posit not just different ideas for the ordering of the same facts, but that they posit also different facts. The 'linguistic relativity principle' seems to point in the same direction. It says, 'in informal terms, that users of markedly different grammars are pointed by their grammars towards different types of observations and different evaluations of externally similar acts of observation, and hence are not equivalent observers, but must arrive at somewhat different views of the world'.[3] But the 'more formal statements'[4] of the principle already contains a different element, for here we are told that 'all observers are not led by *the same physical evidence* to the same picture of the universe, unless their linguistic backgrounds are similar, or can in some way be calibrated',[5] which can either mean that observers using widely different languages will *posit different facts* under the same physical circumstances in the same physical world, or it can mean that they will *arrange similar facts in different ways*. The second interpretation finds some support in the examples given, where different isolates of meaning in English and Shawnee are said to be 'used in reporting *the same experience*'[6] and where we read that 'languages classify items of experience differently';[7] experience is regarded as a uniform reservoir of facts which are *classified* differently by different languages. It finds further support in Whorf's description of the transition from the *horror-vacui* account of barometric phenomena to the modern theory: 'If once these sentences

---

1. Whorf, op. cit., p. 216.
2. Ibid., p. 213.
3. Ibid., p. 221.
4. Ibid., p. 221.
5. Ibid., p. 214, my italics.
6. Ibid., p. 208.
7. Ibid., p. 209.

[Why does water rise in a pump? Because Nature abhors a vacuum.]
seemed satisfying to logic, but today seem idiosyncrasies of a particular
jargon, the change did not come about because science has discovered
new facts. Science has adopted new linguistic formulations of the old
facts, and now that we have become at home in the new dialect, certain
traits of the old one are no longer binding on us'.[8] However, I regard
these more conservative statements as secondary when compared with
the great influence ascribed to grammatical categories and especially to
the more hidden 'rapport systems' of a language.[9]

Whorf and those who follow him regard language as the main and
perhaps as the only 'shaper of events'. That is much too narrow a point of
view. Animals have no language in the sense of Whorf, yet they do not
live in a shapeless world. Planets, at least as conceived today, are not even
alive, but they affect their surroundings and react to them in a lawful
manner. In humans rituals, music, the arts, adaptive behaviour that occurs
without the interposition of words make important contributions to the
way in which the world *appears* and, to those living accordingly, *is*. In the
sciences we have not only statements (the old idea that science is a system
of statements has by now been thoroughly discredited), but observations,
experimental equipment, an intuitive relation between observers and
their equipment that has to be learned in a practical way and cannot be
written down, the work of experimentalists which has much in common
with the work of artists – what they want are not merely results, but
results that emerge in a simple, compelling and aesthetically pleasing
way – and so on. A concentration on language alone, or on 'texts', can
easily lead into absurdity, as is shown by Austin and by the practice of
deconstruction: on the one hand philosophers produce texts, like poets;
on the other hand they take it for granted that their texts reveal a reality
beyond the thoughts, impressions, memories, figures of speech, etc., etc.
from which they arose. (Scientific realists to a certain extent share in this
predicament.)

Finally, some comments on what I think about incommensurability
and how I arrived at the idea.

I think that incommensurability *turns up* when we sharpen our
concepts in the manner demanded by the logical positivists and their

---

8. Ibid., p. 222.
9. Ibid., pp. 68ff.

offspring, and that it *undermines* their ideas on explanation, reduction and progress. Incommensurability *disappears* when we use concepts as scientists use them, in an open, ambiguous and often counter-intuitive manner. Incommensurability is a problem for philosophers, not for scientists, though the latter may become *psychologically* confused by unusual things. I arrived at the phenomenon while studying the early literature on basic statements and by considering the possibility of perceptions radically different from our own. In my thesis[10] I examined the meaning of observational statements. I considered the idea that such statements describe 'what is given' and tried to identify this 'given'. *Phenomenologically* this did not seem to be possible; we notice objects, their properties, their relations, not 'the given'. It is of course true that we can give quick reports on the properties of everyday objects, but this does not change them into non-objects but only shows that we have a special relation to them. Phenomenologically what is given consists of the same things which can also exist unobserved – it is not a new kind of object. Special arrangements such as the reduction screen introduce new conditions, they do not reveal ingredients in objects we already know. Result: the given cannot be isolated by observation.

The second possibility was to isolate it by logical means: what is given can be ascertained *with certainty*, hence I obtain the given contained in the table before me by removing from the statement 'there is a table' all the consequences that make future corrections possible. This shows that the given is the result of an unreasonable decision: untestable statements cannot serve as a basis for science.

Following this argument I introduced the assumption that the meaning of observation statements depends on the nature of the objects described and, as this nature depends on the most advanced theories, on the content of these theories. Or as I formulated it in my first English paper on the topic: the interpretation of an observation language is determined by the theories which we use to explain what we observe, and it changes as soon as these theories change.[11] In a word: observation statements are not just theory-*laden* (the views of Toulmin, Hanson and

---

10. Vienna, 1951 – written after two years of extensive discussion in the Kraft Circle and supervised by Professor Victor Kraft of the University of Vienna.

11. 'An Attempt at a Realistic Interpretation of Experience', *Proc. Arist. Soc.* 1958, reprinted in *Philosophical Papers*, Vol. 1. The passage (in italics) is on p. 31.

apparently also Kuhn) but *fully theoretical* and the distinction between observation statements ('protocol statements' in the terminology of the Vienna Circle) and theoretical statements is a pragmatic distinction, not a semantic distinction; there are no special 'observational meanings'. Thus in the same year as Hanson (Hanson's *Patterns of Discovery* appeared in 1958) and four years before Kuhn I formulated a thesis a weaker form of which became very popular later on. Moreover, my thesis not only was stronger than the thesis of theory-ladenness, it also came from a different source. For while Toulmin and Hanson were inspired by Wittgenstein's *Philosophical Investigations* I started from and returned to ideas that had been developed in the Vienna Circle – and I said so.[12] Quine, whose philosophy shows close connections to the philosophy of the Vienna Circle,[13] also used a criterion of observability that is rather similar to mine.[14]

Now when Feigl heard of these ideas he pointed out that interpreting observations in terms of the theories they are observations of makes nonsense of crucial experiments; for how can an experiment decide between two theories when its interpretation already depends on these theories and when the theories themselves have no common elements, such as a common observation language? In the paper just mentioned and in 'Explanation, Reduction and Empiricism', published in 1962, I took up the challenge. I first increased it by constructing cases where important terms of one theory cannot in any way be defined in another which, moreover, tries to do its job. My example, which I found in Anneliese Maier's *Die Vorläufer Galileis im 14. Jahrhundert* was the relation of the terms 'impetus' and 'momentum'. I also developed a theory of test to answer the challenge. In 1962 I called theories such as those containing 'impetus' and 'momentum' incommensurable theories, said that only a special class of theories, so-called non-instantial theories could be (but need not be) incommensurable and added that successive incommensurable theories are related to each other by replacement, not by subsumption. The year 1962 is also that of Kuhn's great book – but Kuhn used a different approach to apply the same term to a similar

---

12. *Philosophical Papers*, Vol. 1, pp. 49, 125.

13. Details in Dirk Koppelberg, *Die Aufhebung der Analytischen Philosophie*, Frankfurt, 1987.

14. *Philosophical Papers*, Vol. 1, pp. 17f.

(not an identical) situation. His approach was historical, while mine was abstract.

In 1960 I started the studies described in chapters 8, 9 and 16. They revealed that perception and experimentation obey laws of their own which cannot be reduced to theoretical assumptions and are therefore beyond the grasp of theory-bound epistemologies.

I also joined Kuhn in demanding a historical as opposed to an epistemological grounding of science but I still differ from him by opposing the political autonomy of science. Apart from that our views (i.e. my published views and Kuhn's as yet unpublished recent philosophy) by now seem to be almost identical,[15] except that I have little sympathy for Kuhn's attempt to tie up history with philosophical or linguistic, but at any rate with theoretical ropes: a connection with theory just brings us back to what I at least want to escape from – the rigid, though chimaerical (deconstruction!) boundaries of a 'conceptual system'.

15. See my 'Realism and the Historicity of Knowledge', *The Journal of Philosophy*, Vol. LXXXVI, 1989, pp. 353ff, esp. footnote 26 and the postscript to the present essay.

17

Neither science nor rationality are universal measures of excellence. They are particular traditions, unaware of their historical grounding.

So far I have tried to show that reason, at least in the form in which it is defended by logicians, philosophers and some scientists, does not fit science and could not have contributed to its growth. This is a good argument against those who admire science and are also slaves of reason. They must now make a choice. They can keep science; they can keep reason; they cannot keep both.

But science is not sacrosanct. The mere fact that it exists, is admired, has results is not sufficient for making it a measure of excellence. Modern science arose from global objections against earlier views and rationalism itself, the idea that there are general rules and standards for conducting our affairs, affairs of knowledge included, arose from global objections to common sense (example: Xenophanes against Homer). Are we to refrain from engaging in those activities that gave rise to science and rationalism in the first place? Are we to rest content with their results? Are we to assume that everything that happened after Newton (or after Hilbert) is perfection? Or shall we admit that modern science may have basic faults and may be in need of global change? And, having made the admission, how shall we proceed? How shall we localize faults and carry out changes? Don't we need a measure that is independent of science and conflicts with it in order to prepare the change we want to bring about? And will not the rejection of rules and standards that conflict with science forever prevent us from finding such a measure? On the other hand – have not some of the case studies shown that a blunt application of 'rational' procedures would not have given us a better science or a better world, but nothing at all? And how are we to judge the results themselves? Obviously there is no simple way of guiding a practice by rules or of criticizing standards of rationality by a practice.

The problems I have sketched are old ones and much more general than the problem of the relation between science and rationality. They occur whenever a rich, well-articulated and familiar practice – a practice of composing, of painting pictures, of stage production, of selecting people for public office, of keeping order and punishing criminals, a practice of worship, of organizing society – is confronted by a practice of a different kind that can interact with it. The *interactions* and their results depend on historical conditions and vary from one case to the next. A powerful tribe invading a country may impose its laws and change the indigenous traditions by force only to be changed itself by the remnants of the subdued culture. A ruler may decide, for reasons of convenience, to use a popular and stabilizing religion as the basic ideology of his empire and may thereby contribute to the transformation both of his empire and of the religion chosen. An individual, repelled by the theatre of his time and in search of something better, may study foreign plays, ancient and modern theories of drama and, using the actors of a friendly company to put his ideas into practice, change the theatre of a whole nation. A group of painters, desirous of adding the reputation of being scientists to their already enormous reputation as skilled craftsmen, may introduce scientific ingredients such as geometry into painting and thereby create a new style and new problems for painters, sculptors, architects. An astronomer, critical of the difference between classical principles of astronomy and the existing practice and desirous to restore astronomy to its former splendour, may find a way to achieve his aim and so initiate the removal of the classical principles themselves.

In all these cases we have a practice, or a tradition, we have certain influences upon it, emerging from another practice or tradition, and we observe a change. The change may lead to a slight modification of the original practice, it may eliminate it, it may result in a tradition that barely resembles either of the interacting elements.

Interactions such as those just described are accompanied by changing degrees of *awareness* on the part of the participants. Copernicus knew very well what he wanted and so did Constantine the Great (I am now speaking about the initial impulse, not about the transformation that followed). The intrusion of geometry into painting is less easily accounted for in terms of awareness. We have no idea why Giotto tried to achieve a compromise between the surface of the painting and the corporeality of the things painted, especially as pictures were not yet regarded as studies of a material reality. We can surmise that Brunelleschi arrived at his construction by a natural extension of the architects' method of representing three-dimensional objects and that his contacts with contemporary scientists were not without consequence. It is still more difficult to understand the gradually rising claims of artisans to make contributions to the same kind of knowledge whose principles were explained at universities in very different terms. Here we have not a critical *study* of alternative traditions as we have in Copernicus, or in Constantine, but an *impression* of the uselessness of academic science when compared with the fascinating consequences of the journeys of Columbus, Magellan and their successors. There arose then the idea of an 'America of Knowledge', of an entirely new and as yet unforeseen continent of knowledge that could be discovered, just as the real America had been discovered: by a combination of skill and abstract study. Marxists have been fond of confounding insufficient information concerning the awareness that accompanies such processes with irrelevance, and they have ascribed only a secondary role to individual consciousness. In this they were right – but not in the way they thought. For new *ideas*, though often necessary, were not sufficient for explaining the *changes* that now occurred and that depended also on the (often unknown and unrealized) *circumstances* under which the ideas were applied. Revolutions have transformed not only the practices their initiators wanted to change but the very principles by means of which, intentionally or unintentionally, they carried out the change.

Now considering any interaction of traditions we may ask two kinds of questions which I shall call *observer questions* and *participant questions* respectively.

Observer questions are concerned with the details of an interaction. They want to give a historical account of the interaction and, perhaps, formulate laws, or rules of thumb, that apply to all interactions. Hegel's triad: position, negation, synthesis (negation of the negation) is such a rule.

Participant questions deal with the attitude the members of a practice or a tradition are supposed to take towards the (possible) intrusion of another. The observer asks: What happens and what is going to happen? The participant asks: What shall I do? Shall I support the interaction? Shall I oppose it? Or shall I simply forget about it?

In the case of the Copernican Revolution, for example, the observer asks: What impact did Copernicus have on Wittenberg astronomers at about 1560? How did they react to his work? Did they change some of their beliefs, and if so, why? Did their change of opinion have an effect on other astronomers, or were they an isolated group, not taken seriously by the rest of the profession?

The questions of a participant are: This is a strange book indeed – should I take it seriously? Should I study it in detail or only superficially, or should I simply continue as before? The main theses seem absurd at first sight – but, maybe, there is something in them? How shall I find out? And so on.

It is clear that observer questions must take the questions of the participants into account and participants will also listen most carefully (if they are inclined that way, that is) to what observers have to say on the matter – but the intention is different in both cases. Observers want to know what is going on, participants what to do. An observer describes a life he does not lead (except accidentally), a participant wants to arrange his own life and asks himself what attitude to take towards the things that may influence it.

Participants can be *opportunists* and act in a straightforward and practical way. In the late 16th century many princes became Protestants because this furthered their interests, and some of their subjects became Protestants in order to be left in peace. When British colonial officials replaced the laws and habits of foreign tribes and cultures by their own 'civilized' laws the latter were often accepted because they were the laws of the king, or because one had no way to oppose them, and not because of any intrinsic excellence. The source of their power and 'validity' was clearly understood, both by the officials and by the more astute of their unfortunate subjects. In the sciences, and especially in pure mathematics, one often pursues a particular line of research not because it is regarded as intrinsically perfect, but because one wants to see where it leads. I shall call the philosophy underlying such an attitude of a participant a *pragmatic philosophy*.

A pragmatic philosophy can flourish only if the traditions to be judged and the developments to be influenced are seen as temporary makeshifts and not as lasting constituents of thoughts and action. A participant with a pragmatic philosophy views practices and traditions much as a traveller views foreign countries. Each country has features he likes and things he abhors. In deciding to settle down a traveller will have to compare climate, landscape, language, temperament of the inhabitants, possibilities of change, privacy, looks of male and female population, theatre, opportunities for advancement, quality of vices and so on. He will also remember that his initial demands and expectations may not be very sensible and so permit the process of choice to affect and change his 'nature' as well which, after all, is just another (and minor) practice or tradition entering the process. So a pragmatist must be both a participant and an observer even in those extreme cases where he decides to live in accordance with his momentary whims entirely.

Few individuals and groups are pragmatists in the sense just described and one can see why: it is very difficult to see one's own most cherished ideas in perspective, as parts of a changing and, perhaps, absurd tradition. Moreover, this inability not only *exists*, it is also *encouraged* as an attitude proper to those engaged in the study and the improvement of man, society, knowledge. Hardly any religion has ever presented itself just as something worth trying. The claim is much stronger: the religion is the truth, everything else is error and those who know it, understand it but still reject it are rotten to the core (or hopeless idiots).

Two elements are contained in such a claim. First, one distinguishes between traditions, practices and other results of individual and/or collective human activity on the one side and a different domain that may act on the traditions without being one. Secondly, one explains the structure of this special domain in detail. Thus the word of God is powerful and must be obeyed not because the tradition that carries it has much force, but because it is outside all traditions and provides a way of improving them. The word of God can start a tradition, its meaning can be handed on from one generation to the next, but it is itself outside all traditions.

The first element – the belief that some demands are 'objective' and tradition-independent – plays an important role in *rationalism*, which is a secularized form of the belief in the power of the word of God. And this is how the opposition reason–practice obtains its polemical sting. For the two agencies are not seen as two practices which, while perhaps of unequal value, are yet both imperfect and changing human products but as one such product on the one side and lasting measures of excellence on the other. Early Greek rationalism already contains this version of the conflict. Let us examine what circumstances, assumptions, procedures – what features of the historical process – are responsible for it!

To start with the traditions that oppose each other – Homeric common sense and the various forms of rationalism that arise in the 6th to 4th centuries – have *different internal structures*.[1] On the one hand we have complex ideas that cannot be easily explained, they 'work' but one does not know how, they are 'adequate', but one does not know why,

---

1. For details see Chapter 16.

they apply in special circumstances only, are rich in content but poor in similarities and, therefore, in deductive connections. On the other side there are relatively clear and simple concepts which, having just been introduced, reveal a good deal of their structure and can be linked in many ways. They are poor in content, but rich in deductive connections. The difference becomes especially striking in the case of mathematics. In geometry, for example, we start with rules of thumb applying to physical objects and their shapes under a great variety of circumstances. Later on it can be *proved* why a given rule applies to a given case – but the proofs make use of new entities that are nowhere found in nature.

In antiquity the relation between the new entities and the familiar world of common sense gave rise to various theories. One of them which one might call *Platonism* assumes that the new entities are real while the entities of common sense are but their imperfect copies. Another theory, due to the *Sophists*, regards natural objects as real and the objects of mathematics (the objects of 'reason') as simpleminded and unrealistic images of them. These two theories were also applied to the difference between the new and fairly abstract idea of knowledge propagated by Plato (but found already before) and the common-sense knowledge of the time (Plato wisely uses a distorted image of the latter to give substance to the former). Again it was either said that there existed only one true knowledge and that human opinion was but a pale shadow of it or human opinion was regarded as the only substantial knowledge in existence and the abstract knowledge of the philosophers as a useless dream ('I can see horses, Plato,' said Antisthenes, 'but I nowhere see your ideal horse').

It would be interesting to follow this ancient conflict through history down to the present. One would then learn that the conflict turns up in many places and has many shapes. Two examples must suffice to illustrate the great variety of its manifestations.

When Gottsched wanted to reform the German theatre he looked for plays worth imitating. That is, he looked for traditions more orderly, more dignified, more respectable than what he found on the stage of his time. He was attracted by the French theatre and here mainly by Corneille. Being convinced that 'such a complex edifice of poetry (as tragedy) could hardly exist without rules'[2] he looked for the rules and found Aristotle.

---

2. 'Vorrede zum "Sterbenden Cato"' quoted from J. Chr. Gottsched, *Schriften zur Literatur*, Stuttgart, 1972, p. 200.

For him the rules of Aristotle were not a particular way of viewing the theatre, they were the reason for excellence where excellence was found and guides to improvement where improvement seemed necessary. Good theatre was an embodiment of the rules of Aristotle. Lessing gradually prepared a different view. First he restored what he thought to be the real Aristotle as opposed to the Aristotle of Corneille and Gottsched. Next he permitted violations of the letter of Aristotle's rules provided such violations did not lose sight of their aim. And, finally he suggested a different paradigm and emphasized that a mind inventive enough to construct it need not be restricted by rules. If such a mind succeeds in his efforts 'then let us forget the textbook!'[3]

In a different (and much less interesting) domain we have the opposition between those who suggest that languages be constructed and reconstructed in accordance with simple and clear rules and who favourably compare such *ideal languages* with the sloppy and opaque natural idioms and other philosophers who assert that natural languages, being adapted to a wide variety of circumstances, could never be adequately replaced by their anaemic logical competitors.

This tendency to view differences in the structure of traditions (complex and opaque vs simple and clear) as differences in kind (real vs imperfect realization of it) is reinforced by the fact that the critics of a practice take an observer's position with respect to it but remain participants of the practice that provides them with their objections. Speaking the language and using the standards of this practice they 'discover' limitations, faults, errors when all that really happens is that the two practices – the one that is being criticized and the one that does the criticizing – don't fit each other. Many *arguments against* an out-and-out *materialism* are of this kind. They notice that materialism changes the use of 'mental' terms, they illustrate the consequences of the change with amusing absurdities (thoughts having weight and the like) and then they stop. The absurdities show that materialism clashes with our usual ways of speaking about minds, they do not show what is better – materialism or these ways. But taking the participants' point of view with respect to common sense turns

---

3. *Hamburger Dramaturgie*, Stuck 48. Cf., however, Lessing's criticism of the claims of the 'original geniuses' of his time in Stück 96. Lessing's account of the relation between 'reason' and practice is quite complex and in agreement with the view developed further below.

the absurdities into arguments against materialism. It is as if Americans were to object to foreign currencies because they cannot be brought into simple relations (1:1 or 1:10 or 1:100) to the dollar.

The tendency to adopt a participant's view with respect to the position that does the judging and so to create an Archimedian point for criticism is reinforced by certain distinctions that are the pride and joy of armchair philosophers. I refer to the distinction between an evaluation and the fact that an evaluation has been made, a proposal and the fact that the proposal has been accepted, and the related distinction between subjective wishes and objective standards of excellence. When speaking as observers we often say that certain groups accept certain standards, or think highly of these standards. Speaking as participants we equally often *use* the standards without any reference to their origin or to the wishes of those using them. We say 'theories ought to be falsifiable and contradiction free' and not 'I want theories to be falsifiable and contradiction free' or 'scientists become very unhappy unless their theories are falsifiable and contradiction free'. Now it is quite correct that statements of the first kind (proposals, rules, standards) (a) contain no reference to the wishes of individual human beings or to the habits of a tribe and (b) cannot be derived from, or contradicted by, statements concerning such wishes, or habits, or any other facts. But that does not make them 'objective' and independent of traditions. To infer from the absence of terms concerning subjects or groups in 'there ought to be . . .' that the demand made is 'objective' would be just as erroneous as to claim 'objectivity' i.e. independence from personal or group idiosyncrasies, for optical illusions and mass hallucinations on the grounds that the subject, or the group, nowhere occurs in them. There are many statements that are *formulated* 'objectively', i.e. without reference to traditions or practices, but are still *meant to be understood* in relation to a practice. Examples are dates, co-ordinates, statements concerning the value of a currency, statements of logic (after the discovery of alternative logics), statements of geometry (after the discovery of Non-Euclidean geometries) and so on. The fact that the retort to 'you ought to do X' can be 'that's what *you* think!' shows that the same is true of value statements. And those cases where the reply is not allowed can be easily rectified by using discoveries in value theory that correspond to the discovery of alternative geometries, or alternative logical systems: we confront 'objective' value judgement from different cultures or different practices and ask the objectivist how he is going to resolve the conflict.[4] Reduction to shared principles is not always possible and so we must admit that the demands or the formulae expressing them are incomplete as used and have to be revised. Continued insistence on the 'objectivity' of value judgements however would be as illiterate as continued insistence on the 'absolute' use of the pair 'up–down' after discovery of the spherical shape of the earth. And an argument such as 'it is one thing to utter a demand and quite a different thing to assert that a demand has been made – therefore a multiplicity of cultures does not mean relativism' has much in common with the argument that antipodes cannot exist because they would fall 'down'. Both cases rest on antediluvian concepts (and inadequate distinctions). Small wonder our 'rationalists' are fascinated by them.

With this we have also our answer to (b). It is true that stating a demand and describing a practice may be two different things and that logical connections cannot be established between them. This does not mean that the interaction between demands and practices cannot be treated and evaluated as an interaction of practices. For the difference is due, first, to a difference between observer-attitude and participant-attitude: one side, the side defending the 'objectivity' of its values, *uses* its tradition instead of *examining* it – which does not turn the tradition into an objective measure of validity. And secondly, the difference is due to concepts that have been adapted to such one-sidedness. The colonial official who proclaims new laws and a new order in the name of the king has a much better grasp of the situation than the rationalist who recites the mere letter of the law without any reference to the circumstances of its application and who regards this fatal incompleteness as proof of the 'objectivity' of the laws recited.

After this preparation let us now look at what has been called 'the relation between reason and practice'.

---

4. In the play *The Ruling Class* (later turned into a somewhat vapid film with Peter O'Toole) two madmen claiming to be God are confronted with each other. This marvellous idea so confuses the playwright that he uses fire and brimstone instead of dialogue to get over the problem. His final solution, however, is quite interesting. The one madman turns into a good, upright, normal British Citizen who plays Jack the Ripper on the side. Did the playwright mean to say that our modern 'objectivists' who have been through the fire of relativism can return to normalcy only if they are permitted to annihilate all disturbing elements?

Simplifying matters somewhat, we can say that there exist three views on the matter.

A. Reason guides practice. Its authority is independent of the authority of practices and traditions and it shapes the practice in accordance with its demands. This we may call the *idealistic version* of the relation.

B. Reason receives both its content and its authority from practice. It describes the way in which practice works and formulates its underlying principles. This version has been called *naturalism* and it has occasionally been attributed to Hegel (though erroneously so).

Both idealism and naturalism have difficulties.

The difficulties of idealism are that the idealist does not only want to 'act rationally', he also wants his rational actions to have results. And he wants these results to occur not only among the idealizations he uses but in the real world he inhabits. For example, he wants real human beings to build up and maintain the society of his dreams, he wants to understand the motions and the nature of real stars and real stones. Though he may advise us to 'put aside (all observation of) the heavens'[5] and to concentrate on ideas only he eventually returns to nature in order to see to what extent he has grasped its laws.[6] It then often turns out and it often has turned out that acting rationally in the sense preferred by him does not produce the expected results. This conflict between rationality and expectations was one of the main reasons for the constant reform of the canons of rationality and much encouraged naturalism.

But naturalism is not satisfactory either. Having chosen a popular and successful practice the naturalist has the advantage of 'being on the right side', at least for the time being. But a practice may deteriorate; or it may be popular for the wrong reasons. (Much of the popularity of modern scientific medicine is due to the fact that sick people have nowhere else to go and that television, rumours, the technical circus of well equipped hospitals convince them that they could not possibly do better.) Basing standards on a practice and leaving it at that may forever perpetuate the shortcomings of this practice.

The difficulties of naturalism and idealism have certain elements in common. The inadequacy of standards often becomes clear from the barrenness of the practice they engender, the shortcomings of practices often are very obvious when practices based on different standards

---

5. Plato, *Republic*, 530bf.
6. *Epinomis.*

flourish. This suggests that reason and practice are not two different kinds of entities but *parts of a single dialectical process.*

The suggestion can be illustrated by the relation between a map and the adventures of a person using it or by the relation between an artisan and his instruments. Originally maps were constructed as images of and guides to reality and so, presumably, was reason. But maps, like reason, contain idealizations (Hecataeus of Miletus, for example, imposed the general outlines of Anaximander's cosmology on his account of the occupied world and represented continents by geometrical figures). The wanderer uses the map to find his way but he also corrects it as he proceeds, removing old idealizations and introducing new ones. Using the map no matter what will soon get him into trouble. But it is better to have maps than to proceed without them. In the same way, the example says, reason without the guidance of a practice will lead us astray while a practice is vastly improved by the addition of reason.

This account, though better than naturalism and idealism and much more realistic, is still not entirely satisfactory. It replaces a one-sided action (of reason upon practice or practice upon reason) by an interaction but it retains (certain aspects of) the old views of the interacting agencies: reason and practice are still regarded as entities of different kinds. They are both needed but reason can exist without a practice and practice can exist without reason. Shall we accept this account of the matter?

To answer the question we need only remember that the difference between 'reason' and something 'unreasonable' that must be formed by it or can be used to put it in its place arose from turning structural differences of practices into differences of kind. Even the most perfect standards or rules are not independent of the material on which they act (how else could they find a point of attack in it?) and we would hardly understand them or know how to use them were they not well-integrated parts of a rather complex and in places quite opaque practice or tradition, viz. the language in which the *defensor rationis* expresses his stern commands.[7] On the other hand even the most disorderly practice

---

7. This point has been made with great force and with the help of many examples by Wittgenstein (see my essay 'Wittgenstein's *Philosophical Investigations*', *Phil Rev.*, 1955). What have rationalists replied? Russell (coldly): 'I don't understand.' Sir Karl Popper (breathlessly): 'He is right, he is right – I don't understand it either!' In a word: the point is irrelevant because leading rationalists don't understand it. I, on the other hand, would start doubting the intelligence (and perhaps also the intellectual honesty) of rationalists who don't understand (or pretend not to understand) such a simple point.

is not without its regularities, as emerges from our attitude towards non-participants.[8] *What is called 'reason' and 'practice' are therefore two different types of practice*, the difference being that the one clearly exhibits some simple and easily producible formal aspects, thus making us forget the complex and hardly understood properties that guarantee the simplicity and producibility, while the other drowns the formal aspects under a great variety of accidental properties. But complex and implicit reason is still reason, and a practice with simple formal features hovering above a pervasive but unnoticed background of linguistic habits is still a practice. Disregarding (or, rather, not even noticing) the sense-giving and application-guaranteeing mechanism in the first case and the implicit regularities in the second a rationalist perceives law and order here and material yet in need of being shaped there. The habit, also commented upon in an earlier part of this section, to take a participant's point of view with respect to the former and an observer's attitude towards the latter further separates what is so intimately connected in reality. And so we have finally two agencies, stern and orderly reason on the one side, a malleable but not entirely yielding material on the other, and with this all the 'problems of rationality' that have provided philosophers with intellectual (and, let us not forget, also with financial) nourishment ever since the 'Rise of Rationalism in the West'. One cannot help noticing that the arguments that are still used to support this magnificent result are indistinguishable from those of the theologian who infers a creator wherever he sees some kind of order: obviously order is not inherent in matter and so must have been imposed from the outside.

The interaction view must therefore be supplemented with a satisfactory account of the interacting agencies. Presented in this way it becomes a triviality. For there is no tradition no matter how hard-headed its scholars and how hard-limbed its warriors that will remain unaffected by what occurs around it. At any rate – what changes, and how, is now a matter either for *historical research* or for *political action* carried out by those who participate in the interacting traditions.

I shall now state the implications of these results in a series of theses with corresponding explanations.

---

8. See my short comments on 'covert classifications' in Chapter 16.

We have seen that rational standards and the arguments supporting them are visible parts of special traditions consisting of clear and explicit principles and an unnoticed and largely unknown but absolutely necessary background of dispositions for action and judgement. The standards become 'objective' measures of excellence when adopted by participants of traditions of this kind. We have then 'objective' rational standards and arguments for their validity. We have further seen that there are other traditions that also lead to judgements though not on the basis of explicit standards and principles. These value judgements have a more 'immediate' character, but they are still evaluations, just like those of the rationalist. In both cases judgements are made by individuals who participate in traditions and use them to separate 'Good' from 'Evil'. We can therefore state:

i. *Traditions are neither good nor bad, they simply are.* 'Objectively speaking', i.e. independently of participation in a tradition, there is not much to choose between humanitarianism and anti-Semitism.

Corollary: rationality is not an arbiter of traditions, it is itself a tradition or an aspect of a tradition. It is therefore neither good nor bad, it simply is.

ii. *A tradition assumes desirable or undesirable properties only when compared with some tradition,* i.e. only when viewed by participants who see the world in terms of its values. The projections of these participants *appear objective* and statements describing them *sound objective* because the participants and the tradition they project are nowhere mentioned in them. They *are subjective* because they depend on the tradition chosen and on the use the participants make of it. The subjectivity is noticed as soon as participants realize that different traditions give rise to different judgements. They will then have to revise the content of their value statements just as physicists revised the content of even the simplest statement concerning length when it was discovered that length depends on reference systems and just as everybody revised the content of 'down' when it was discovered that the earth is spherical. Those who don't carry out the revision cannot pride themselves on forming a special school of especially astute philosophers who have overcome moral relativism, just as those who still cling to absolute lengths cannot pride themselves on forming a special school of especially astute physicists who have overcome relativity. They are just pig-headed, or badly informed, or both.

iii. *i. and ii. imply a relativism of precisely the kind that seems to have been defended by Protagoras.* Protagorean relativism is *reasonable* because it pays attention to the pluralism of traditions and values. And it is *civilized* for it does not assume that one's own village and the strange customs it contains are the navel of the world.[9]

iv. *Every tradition has special ways of gaining followers.* Some traditions reflect about these ways and change them from one group to the next. Others take it for granted that there is only one way of making people accept their views. Depending on the tradition adopted this way will look acceptable, laughable, rational, foolish, or will be pushed aside as 'mere propaganda'. Argument is propaganda for one observer, the essence of human discourse for another.

v. We have seen that individuals or groups participating in the interaction of traditions may adopt a pragmatic philosophy when judging the events and structures that arise. The principles of their philosophy often emerge only during the interaction (people change while observing change or participating in it and the traditions they use may change with them). This means that *judging a historical process one may use an as yet unspecified and unspecifiable practice.* One may base judgements and actions on standards that cannot be specified in advance but are introduced by the very judgements (actions) they are supposed to guide and one may even act without any standards, simply following some natural inclination. The fierce warrior who cures his wounded enemy instead of killing him has no idea why he acts as he does and gives an entirely erroneous account of his reasons. But his action introduces an age of collaboration and peaceful competition instead of permanent hostility and so creates a new tradition of commerce between nations. The question – How will you decide what path to choose? How will you know what pleases you and what you want to reject? – has therefore at least two answers, viz. (1) there is no decision but a natural development leading to traditions which in retrospect give reasons for the action had it been a decision in accordance with standards or (2) to ask how one will judge and choose in as yet unknown surroundings makes as much sense as to ask what measuring instruments one will use in as yet unexplored domains. Standards which are intellectual measuring instruments often

---

9. Protagoras is discussed in detail in Chapter 1, sections 3ff of *Farewell to Reason*.

have to be *invented* to make sense of new historical situations just as measuring instruments have constantly to be invented to make sense of new physical situations.

vi. There are therefore at least *two different ways of collectively deciding an issue* which I shall call a *guided exchange* and an *open exchange* respectively.

In the first case some or all participants adopt a well-specified tradition and accept only those responses that correspond to its standards. If one party has not yet become a participant of the chosen tradition he will be badgered, persuaded, 'educated' until he does – and then the exchange begins. Education is separated from decisive debates, it occurs at an early stage and guarantees that the grown-ups will behave properly. A *rational debate* is a special case of a guided exchange. If the participants are rationalists then all is well and the debate can start right away. If only some participants are rationalists and if they have power (an important consideration!) then they will not take their collaborators seriously until they have also become rationalists: a society based on rationality is not entirely free; one has to play the game of the intellectuals.[10]

An open exchange, on the other hand, is guided by a pragmatic philosophy. The tradition adopted by the parties is unspecified in the beginning and develops as the exchange proceeds. The participants get immersed into each other's ways of thinking, feeling, perceiving to such an extent that their ideas, perceptions, world-views may be entirely changed – they become different people participating in a new and different tradition. An open exchange respects the partner whether he is an individual or an entire culture, while a rational exchange promises respect only within the framework of a rational debate. An open exchange has no organon though it may invent one, there is no logic though new forms of logic may emerge in its course. An open exchange establishes connections between different traditions and transcends the relativism of points iii and iv. However, it transcends it in a way that cannot be made objective but depends in an unforeseeable manner on the (historical,

---

10. 'It is perhaps hardly necessary to say', says John Stuart Mill, 'that this doctrine (pluralism of ideas and institutions) is meant to apply only to human beings in the maturity of their faculties' – i.e. to fellow intellectuals and their pupils. 'On Liberty', in *The Philosophy of John Stuart Mill*, ed. M. Cohen, New York, 1961, p. 197.

psychological, material) conditions in which it occurs. (See also the last paragraph of Chapter 16.)

vii. *A free society is a society in which all traditions are given equal rights, equal access to education and other positions of power.* This is an obvious consequence of i, ii and iii. If traditions have advantages only from the point of view of other traditions then choosing one tradition as a basis of a free society is an arbitrary act that can be justified only by resorting to power. A free society thus cannot be based on any particular creed; for example, it cannot be based on rationalism or on humanitarian considerations. The basic structure of a free society is a *protective structure*, not an ideology, it functions like an iron railing not like a conviction. But how is this structure to be conceived? Is it not necessary to *debate* the matter or should the structure be simply *imposed*? And if it is necessary to debate the matter then should this debate not be kept free from subjective influences and based on 'objective' considerations only? This is how intellectuals try to convince their fellow citizens that the money paid to them is well spent and that their ideology should continue to assume the central position it now has. I have already exposed the errors-cum-deceptions behind the phrase of the 'objectivity of a rational debate': the standards of such a debate *are not* 'objective' they only *appear to be* 'objective' because reference to the group that profits from their use has been omitted. They are like the invitations of a clever tyrant who instead of saying 'I want you to do . . .' or 'I and my wife want you to do . . .' says 'What all of us want is . . .' or 'What the gods want of us is . . .' or, even better, 'It is rational to do . . .' and so seems to leave out his own person entirely. It is somewhat depressing to see how many intelligent people have fallen for such a shallow trick. We remove it by observing:

viii. that *a free society will not be imposed but will emerge only where people engaging in an open exchange* (cf. vi above) *introduce protective structures of the kind alluded to.* Citizen initiatives on a small scale, collaboration between nations on a larger scale are the developments I have in mind. The United States are not a free society in the sense described here.

ix. *The debates settling the structure of a free society are open debates not guided debates.* This does not mean that the concrete developments described under the last thesis *already use* open debates, it means that they *could use* them and that rationalism is not a necessary ingredient of the basic structure of a free society.

The results for science are obvious. Here we have a particular tradition, 'objectively' on par with all other traditions (theses i and vii). Its results will appear magnificent to some traditions, execrable to others, barely worth a yawn to still further traditions. Of course, our well-conditioned materialistic contemporaries are liable to burst with excitement over events such as the moonshots, the double helix, non-equilibrium thermodynamics. But let us look at the matter from a different point of view, and it becomes a ridiculous exercise in futility. It needed billions of dollars, thousands of well-trained assistants, years of hard work to enable some inarticulate and rather limited contemporaries[11] to perform a few graceless hops in a place nobody in his right mind would think of visiting – a dried out, airless, hot stone. But mystics, using only their minds, travelled across the celestial spheres to God himself, whom they viewed in all his splendour, receiving strength for continuing their lives and enlightenment for themselves and their fellow men. It is only the illiteracy of the general public and of their stern trainers, the intellectuals, and their amazing lack of imagination that makes them reject such comparisons without further ado. A free society does not object to such an attitude but it will not permit it to become a basic ideology either.

x. *A free society insists on the separation of science and society.* More about this topic in Chapter 19.

---

11. See Norman Mailer, *Of a Fire on the Moon*, London, 1970.

Yet it is possible to evaluate standards of rationality and to improve them. The principles of improvement are neither above tradition nor beyond change and it is impossible to nail them down.

I shall now illustrate some of these results by showing how standards are and have been criticized in physics and astronomy and how this procedure can be extended to other fields.

Chapter 17 started with the general problem of the relation between reason and practice. In the illustration reason becomes scientific rationality, practice the practice of scientific research, and the problem is the relation between scientific rationality and research. I shall discuss the answers given by idealism, naturalism and by a third position, not yet mentioned, which I shall call naive anarchism.

According to *idealism* it is rational (proper, in accordance with the will of the gods – or whatever other encouraging words are being used to befuddle the natives) to do certain things – *come what may*. It is rational (proper, etc.) to kill the enemies of the faith, to avoid *ad hoc* hypotheses, to despise the desires of the body, to remove inconsistencies, to support progressive research programmes and so on. Rationality (justice, the Divine Law) are universal, independent of mood, context, historical circumstances and give rise to equally universal rules and standards.

There is a version of idealism that seems to be somewhat more sophisticated but actually is not. Rationality (the law, etc.) is no longer said to be universal, but there are universally valid conditional statements asserting what is rational in what context and there are corresponding conditional rules.

Some reviewers have classified me as an idealist in the sense just described with the proviso that I try to replace familiar rules and standards by more 'revolutionary' rules such as proliferation and counterinduction and almost everyone has ascribed to me a 'methodology' with 'anything goes' as its one 'basic principle'. But in Chapter 21 I say quite explicitly that 'my intention is not to replace one set of general rules by another such set: my intention is, rather, to convince the reader that, *all methodologies, even the most obvious ones, have their limits*' or, to express it in terms just explained, my intention is to show that idealism, whether of the simple or of the context-dependent kind, is the wrong solution for the problems of scientific rationality. These problems are not solved by a change of standards but by taking a different view of standards altogether.

Idealism can be dogmatic and it can be critical. In the first case the rules proposed are regarded as final and unchangeable; in the second case there is the possibility of discussion and change. But the discussion does not take practices into account – it remains restricted to an abstract domain of standards, rules and logic.

The limitation of all rules and standards is recognized by *naive anarchism*. A naive anarchist says (a) that both absolute rules and context-dependent rules have their limits and infers (b) that all rules and standards are worthless and should be given up. Most reviewers regard me as a naive anarchist in this sense, overlooking the many passages where I show how certain procedures *aided* scientists in their research. For in my studies of Galileo, of Brownian motion, of the Presocratics I not only demonstrate the *failures* of familiar standards, I also try to show what not so familiar procedures did actually *succeed*. Thus while I agree with (a) I do not agree with (b). I argue that all rules have their limits and that there is no comprehensive 'rationality', I do not argue that we should proceed without rules and standards. I also argue for a contextual account but again the contextual rules are not to *replace* the absolute rules, they are to *supplement* them. Moreover, I suggest a new *relation* between rules and practices. It is this relation and not any particular rule-content that characterizes the position I wish to defend.

This position adopts some elements of *naturalism* but it rejects the naturalist philosophy. According to naturalism rules and standards are obtained by an analysis of traditions. As we have seen, the problem is which tradition to choose. Philosophers of science will of course opt for science as their basic tradition. But science is not *one* tradition, it is *many*, and so it gives rise to many and partly incompatible standards (I have explained this difficulty in my discussion of Lakatos).[1] Besides, the procedure makes it impossible for the philosopher to give reasons for his choice of science over myth or Aristotle. Naturalism cannot solve the problem of scientific rationality.

As in Chapter 17 we can now compare the drawbacks of naturalism and idealism and arrive at a more satisfactory view. Naturalism says that reason is completely *determined* by research. Of this we retain

---

1. *Philosophical Papers*, Vol. 2, Chapter 10. See also Chapter 19.

the idea that research can change reason. Idealism says that reason completely *governs* research. Of this we retain the idea that reason can change research. Combining the two elements we arrive at the idea of *a guide who is part of the activity guided and is changed by it*. This corresponds to the interactionist view of reason and practice formulated in Chapter 17 and illustrated by the example of the map. Now the interactionist view assumes two different entities, a disembodied guide on the one side and a well-endowed practice on the other. But the guide seems disembodied only because its 'body', i.e. the very substantial practice that underlies it, is not noticed and the 'practice' seems crude and in need of a guide only because one is not aware of the complex and rather sophisticated laws it contains. Thus the problem is not the interaction of a practice with something different and external, *but the development of one tradition under the impact of others*. A look at the way in which science treats its problems and revises its 'standards' confirms this picture.

In physics theories are used both as descriptions of facts and as standards of speculation and factual accuracy. *Measuring instruments* are constructed in accordance with laws and their readings are tested under the assumption that these laws are correct. In a similar way theories giving rise to physical principles provide standards to judge other *theories* by: theories that are relativistically invariant are better than theories that are not. Such standards are of course not untouchable. The standard of relativistic invariance, for example, may be removed when one discovers that the theory of relativity has serious shortcomings. Shortcomings are occasionally found by a direct examination of the theory, for example by an examination of its mathematics, or its predictive success. They may also be found by the development of alternatives (see Chapter 3) – i.e. by research that violates the standards to be examined.

The idea that nature is infinitely rich both qualitatively and quantitatively leads to the desire to make new discoveries and thus to a principle of content increase which gives us another standard to judge theories by: theories that have excess content over what is already known are preferable to theories that have not. Again the standard is not untouchable. It is in trouble the moment we discover that we inhabit a finite world. The discovery is prepared by the development of 'Aristotelian' theories which refrain from going beyond a given set of properties – it is again prepared by research that violates the standard.

The procedure used in both cases contains a variety of elements and so there are different ways of describing it, or reacting to it.

One element and to my mind the most important one is *cosmological*. The standards we use and the rules we recommend make sense only in a world that has a certain structure. They become inapplicable, or start running idle in a domain that does not exhibit this structure. When people heard of the new discoveries of Columbus, Magellan, Diaz they realized that there were continents, climates, races not enumerated in the ancient accounts and they conjectured there might be new continents of knowledge as well, that there might be an 'America of Knowledge' just as there was a new geographical entity called 'America', and they tried to discover it by venturing beyond the limits of the received ideas. The demand for content increase now became very plausible. It arose from the wish to discover more and more of a nature that seemed to be infinitely rich in extent and quality. The demand has no point in a finite world that is composed of a finite number of basic qualities.

How do we find the cosmology that supports or suspends our standards? The reply introduces the second element that enters the revision of standards, viz. *theorizing* in a general sense, including myth and metaphysical speculation. The idea of a finite world becomes acceptable when we have theories describing such a world and when these theories turn out to be better than their infinitist rivals. The world is not directly given to us, we have to catch it through the medium of traditions which means that even the cosmological argument refers to a certain stage of competition between world-views, theories of rationality included.

Now when scientists become accustomed to treating theories in a certain way, when they forget the reasons for this treatment but simply regard it as the 'essence of science' or as an 'important part of what it means to be scientific', when philosophers aid them in their forgetfulness by systematizing the familiar procedures and showing how they flow from an abstract theory of rationality, then the theories needed to show the shortcomings of the underlying standards will not be introduced or, if they are introduced, will not be taken seriously. They will not be taken seriously because they clash with customary habits and systematizations thereof.

For example, a good way of examining the idea that the world is finite both qualitatively and quantitatively is to develop an Aristotelian cosmology. Such a cosmology provides means of description adapted to a finite world while the corresponding methodology replaces the demand for content increase by the demand for adequate descriptions of this kind. Assume we introduce theories that correspond to the cosmology and develop them in accordance with the new rules. What will happen? Scientists will be unhappy for the theories have unfamiliar properties. Philosophers of science will be unhappy because they introduce standards unheard of in their profession. Being fond of surrounding their unhappiness with arias called 'reasons' they will go a little further. They will say that they are not merely unhappy, but have 'arguments' for their unhappiness. The arguments in most cases are elaborate repetitions and variations of the standards they grew up with and so their cognitive content is that of 'But the theory is ad hoc!' or 'But the theories are developed without content increase!' And all one hears when asking the further question why that is so bad is either that science has proceeded differently for at least 200 years or that content increase solves some problems of confirmation theory. Yet the question was not what science does but how it can be improved and whether adopting some confirmation theories is a good way of learning about the world. No answer is forthcoming. And so interesting possibilities are removed by firmly insisting on the status quo. It is amusing to see that such insistence becomes the more determined the more 'critical' the philosophy that is faced with the problem. We, on the other hand, retain the lesson that *the validity, usefulness, adequacy of popular standards can be checked only by research that violates them.*

A further example, to illustrate the point. The idea that information concerning the external world travels undisturbed via the senses into the mind leads to the standard that all knowledge must be checked by observation: theories that agree with observation are preferable to theories that do not. This simple standard is in need of replacement the moment we discover that sensory information is distorted in many ways. We make the discovery when developing theories that conflict with observation and finding that they excel in many other respects (Chapters 5 to 11 describe how Galileo contributed to the discovery).

Finally, the idea that things are well defined and that we do not live in a paradoxical world leads to the standard that our knowledge must be self-consistent. Theories that contain contradictions cannot be part of science. This apparently quite fundamental standard which many philosophers accept as unhesitatingly as Catholics once accepted the dogma of the immaculate conception of the Virgin loses its authority the moment we find that there are facts whose only adequate description is inconsistent and that inconsistent theories may be fruitful and easy to handle while the attempt to make them conform to the demands of consistency creates useless and unwieldy monsters.[2]

The last example raises further questions which are usually formulated as objections against it (and against the criticism of other standards as well, standards of content increase included).

One objection is that non-contradiction is a necessary condition of research. A procedure not in agreement with this standard is not research – it is chaos. It is therefore not possible to examine non-contradiction in the manner described in the last example.

The main part of the objection is the second statement and it is usually supported by the remark that a contradiction implies every statement. This it does – but only in rather simple logical systems. Now it is clear that changing standards or basic theories has repercussions that must be taken care of. Admitting velocities larger than the velocity of light into relativity and leaving everything else unchanged gives us some rather puzzling results such as imaginary masses and velocities. Admitting well-defined positions and momenta into the quantum theory and leaving everything else unchanged creates havoc with the laws of interference. Admitting contradictions into a system of ideas allegedly connected by the laws of standard logic and leaving everything else unchanged makes us assert every statement. Obviously we shall have to make some further changes, for example we shall have to change some rules of derivation in the last case. Carrying out the change removes the problems and research can proceed as planned. (Scientific practice containing inconsistencies is already arranged in the right way.)

But – says an objection that is frequently raised at this point – How will the results of the research be evaluated if fundamental standards have been removed? For example, what standards show that research in violation of content increase produces theories which are '*better* than their infinitist rivals' as I said a few paragraphs ago? Or what standards show that theories in conflict with observations have something to offer while their observationally impeccable rivals have not? Does not a decision to accept unusual theories and to reject familiar ones assume standards

---

2. See Chapter 16, text to footnotes 91ff.

and is it not clear, therefore, that cosmological investigations cannot try to provide alternatives to all standards? These are some of the questions one hears with tiring regularity in the discussion of 'fundamental principles' such as consistency, content increase, observational adequacy, falsifiability, and so on. It is not difficult to answer them.

It is asked how research leading to the revision of standards is to be evaluated. For example, when and on what grounds shall we be satisfied that research containing inconsistencies has revealed a fatal shortcoming of the standard of non-contradiction? The question makes as little sense as the question of what measuring instruments will help us to explore an as-yet-unspecified region of the universe. We don't know the region, we cannot say what will work in it. To advance we must either enter the region, or start making conjectures about it. We enter the region by articulating unusual intellectual, social, emotional tendencies, no matter how strange they may seem when viewed through the spectacles of established theories or standards. It would certainly be silly to disregard *physical features* that do not agree with deeply ingrained spiritual notions. But it is equally shortsighted to curtail *fantasies* that do not seem to fit into the physical universe. Fantasies and, in fact, the entire subjectivity of human beings are just as much a part of the world as fleas, stones and quarks and there is no reason why we should change them to protect the latter.

Similar considerations apply to the standards that are supposed to guide our thoughts and actions. They are not stable and they cannot be stabilized by tying them to a particular point of view. For Aristotle knowledge was qualitative and observational. Today knowledge is quantitative and theoretical, at least as far as our leading natural scientists are concerned. Who is right? That depends on what kind of information has privileged status and this in turn depends on the culture, or the 'cultural leaders' who use the information. Many people, without much thought, prefer technology to harmony with Nature; hence, quantitative and theoretical information is regarded as 'real' and qualities as 'apparent' and secondary. But a culture that centres on humans, prefers personal acquaintance to abstract relations (intelligence quotients; efficiency statistics) and a naturalist's approach to that of molecular biologists will say that knowledge is qualitative and will interpret quantitative laws as bookkeeping devices, not as elements of reality.

Combining the considerations of the last two paragraphs we see that even the apparently hardest scientific 'fact' can be dissolved by decisions undermining the values that make it a fact and/or by research that replaces it by facts of a different kind. This is not a new procedure. Philosophers from Parmenides to 20th-century (undialectical) materialists and scientists from Galileo and Descartes to Monod used it to devalue, and to declare as mere appearance, the qualitative features of human life. But what can be used to support science can also be used against it. The (cultural) measuring instruments that separate 'reality' from 'appearance' change and must change when we move from one culture to another and from one historical stage to the next, just as our physical measuring instruments change and must change when we leave one physical region (one historical period) and enter another.

Science is neither a single tradition, nor the best tradition there is, except for people who have become accustomed to its presence, its benefits and its disadvantages. In a democracy it should be separated from the state just as churches are now separated from the state.

I shall now summarize the arguments of the preceding chapters by trying to answer the following three questions.

1. *What is science?* How do scientists proceed, how do their standards differ from the standards of other enterprises?

2. *What's so great about science?* What are the reasons that might compel us to prefer the sciences to other forms of life and ways of gathering knowledge?

3. *How are we to use the sciences and who decides the matter?*

My answer to the first question is that the wide divergence of individuals, schools, historical periods, entire sciences makes it extremely difficult to identify comprehensive principles either of method, or of fact. The word 'science' may be a single word – but there is no single entity that corresponds to that word.

In the domain of *method* we have scientists like Salvador Luria who want to tie research to events permitting 'strong inferences', 'predictions that will be strongly supported and sharply rejected by a clear-cut experimental step'.[1]

According to Luria the experiments (Luria and Delbrueck, 1943) which showed that the resistance of bacteria to phage invasion is a result of environment-independent mutations and not of an adaptation to the environment had precisely this character. There was a simple prediction: fluctuations, from one culture to the next, of surviving colonies of bacteria on an agar containing an excess of bacteriophages would be small in the first case, but would contain avalanches in the second. The prediction could be tested in a simple and straightforward way and there was a decisive result. (The result refuted Lamarckism, which was popular among bacteriologists but practically extinct elsewhere – a first indication of the complexity of science.)

---

1. S.E. Luria, *A Slot Machine, a Broken Test Tube*, New York, 1985, p. 115.

Scientists inclined in the manner of Luria show a considerable 'lack of enthusiasm in the "big problems" of the Universe or of the early Earth or in the concentration of carbon dioxide in the upper atmosphere',[2] all subjects that are 'loaded with weak inferences'.[3] In a way they are continuing the Aristotelian approach which demands close contact with experience and objects to following a plausible idea to the bitter end.[4]

However, this was precisely the procedure adopted by Einstein, by researchers in celestial mechanics between Newton and Poincaré, by the proponents of atomism and, later, the kinetic theory, by Heisenberg during the initial stages of matrix mechanics and by almost all cosmologists. Einstein's first cosmological paper is a purely theoretical exercise containing not a single astronomical constant. The subject of cosmology itself for a long time found few supporters among physicists. Hubble the observer was respected, the rest had a hard time:

> Journals accepted papers from observers, giving them only the most cursory refereeing whereas our own papers always had a stiff passage, to a point where one became quite worn out with explaining points of mathematics, physics, fact and logic to the obtuse minds who constitute the mysterious anonymous class of referees, doing their work, like owls, in the darkness of the night.[5]

'Is it not really strange', asks Einstein, 'that human beings are normally deaf to the strongest argument while they are always inclined to overestimate measuring accuracies?'[6] – but just such an 'overestimating of measuring accuracies' is the rule in epidemiology, demography, genetics, spectroscopy and in other subjects. The variety increases when we move into sciences like cultural anthropology, where a compromise has to be found between the effects of personal contact and the idea of an objective approach on the one side and the practical needs for quick

---

2. Ibid., p. 119.

3. Ibid.

4. *De Coelo* 293a24ff.

5. F. Hoyle in Y. Terzian and E.M. Bilson (eds), *Cosmology and Astrophysics*, Ithaca and London, 1982, p. 21.

6. Letter to Max Born, quoted from the *Born–Einstein Letters*, New York, 1971, p. 192.

action and theoretical thoroughness on the other. 'To hear a seminar at a university about modes of production in the morning', writes Robert Chambers,

> and then attend a meeting in a government office about agricultural extension in the afternoon leaves a schizoid feeling. One might not know that both referred to the same small farmers and might doubt whether either discussion has anything to contribute to the other.[7]

But is it not true that scientists proceed in a methodical way, avoid accidents and pay attention to observation and experiment? Not always. Some scientists propose theories and calculate cases which have little or no connection with reality. 'The great growth in technical achievements which began in the nineteenth century', we read in L. Prandtl's lectures *Fundamentals of Hydro- and Aeromechanics*,

> left scientific knowledge far behind. The multitudinous problems of practice could not be answered by the hydrodynamics of Euler; they could not even be discussed. This was chiefly because, starting from Euler's equations of motion the science had become more and more a purely academic analysis of the hypothetical frictionless 'ideal fluid'. This theoretical development is associated with the names of Helmholtz, Kelvin, Lamb and Rayleigh.
>
> The analytical results obtained by means of this so called 'classical hydrodynamics' virtually do not agree at all with the practical phenomena. ... Therefore the engineers ... put their trust in a mass of empirical data collectively known as the 'science of hydraulics', a branch of knowledge which grew more and more unlike hydrodynamics.[8]

According to Prandtl we have a disorderly collection of facts on the one side, sets of theories starting from simple but counterfactual assumptions on the other, and no connection between the two. More recently the axiomatic approach in quantum mechanics and especially in quantum field theory was compared by cynical observers to the shakers, 'a religious sect of New England who built solid barns and led celibate lives, a non-

---

7. *Rural Development*, London, 1983, p. 29.
8. Ed. O.G. Tietjens, New York, 1954, p. 3.

scientific equivalent of proving rigorous theorems and calculating no cross sections'.[9]

Yet in quantum mechanics this apparently useless activity has led to a more coherent and far more satisfactory codification of the facts than had been achieved before, while in hydrodynamics 'physical commonsense' occasionally turned out to be less accurate than the results of rigorous proofs based on wildly unrealistic assumptions. An early example is Maxwell's calculation of the viscosity of gases. For Maxwell this was an exercise in theoretical mechanics, an extension of his work on the rings of Saturn. Neither he nor his contemporaries believed the outcome – that viscosity remains constant over a wide range of density – and there was contrary evidence. Yet more precise measurements confirmed the prediction.[10] Few people were prepared for such a turn of events. Mathematical curiosity had started the work, cross-fertilization, not general principles, had brought it to a conclusion.

Meanwhile the situation has changed in favour of theory. In the sixties and seventies, when science was still in public favour, theory got the upper hand, at universities, where it increasingly replaced professional skills, even in medicine, and in special subjects such as biology or chemistry where earlier morphological and substance-related research was replaced by a study of molecules. In cosmology a firm belief in the Big Bang now tends to devalue observations that clash with it. 'Such observations', writes C. Burbidge,

> are delayed at the refereeing stage as long as possible with the hope that the author will give up. If this does not occur and they are published the second line of defence is to ignore them. If they give rise to some

---

9. R.F. Streater and A.S. Wightman, *PCT, Spin, Statistics and All That*, New York, 1964, p. 1.

10. For quantum mechanics see sections 4.1 and 4.2 of Hans Primas, *Chemistry, Quantum Mechanics and Reductionism*, Berlin–New York, 1981. Maxwell's calculations are reproduced in *The Scientific Papers of James Clerk Maxwell*, ed. W.D. Niven, New York, 1965 (first published in 1890), pp. 377ff. The conclusion is stated on p. 391: 'A remarkable result here presented to us . . . is that if this explanation of gaseous friction be true, the coefficient of friction is independent of the density. Such a consequence of a mathematical theory is very startling, and the only experiment I have met with on the subject does not seem to confirm it.' For examples from hydrodynamics see G. Birkhoff, *Hydrodynamics*, New York, 1955, sections 20 and 21.

comment, the best approach is to argue simply that they are hopelessly wrong and then, if all else fails, an observer may be threatened with loss of telescope time until he changes his program.[11]

Thus all we can say is that scientists proceed in many different ways, that rules of method, if mentioned explicitly, are either not obeyed at all, or function at most like rules of thumb and that important results come from the confluence of achievements produced by separate and often conflicting trends. The idea that ' "scientific" knowledge is in some way peculiarly positive and free from differences of opinion'[12] is nothing but a chimaera.

The situation in the arts is quite similar – as a matter of fact, it occurs in all areas of human activity. Cennino Cennini's *Libro dell'Arte* of 1390 contains practical advice based on a rich experience and complex skills. Leon Battista Alberti's *Della Pittura* of 1435/6 is a theoretical treatise closely tied to central perspective and academic optical theory. Perspective soon became a mania among artists. Leonardo and Raphael then pointed out, the one in words, the other practically (see the sphere on the right-hand side of his *School of Athens* in the Stanza della Segnatura of the Vatican), that a picture that is to be viewed under normal circumstances, from a comfortable but not well-defined distance and with both eyes wide open cannot obey the rules of central perspective. They thereby clarified the difference between physiological optics and geometrical optics which Kepler, more than a century later, still tried to bridge by an easily refuted hypothesis (see Ch. 9, text to footnote 50). But central perspective remained a basis on which various changes were superimposed.

So far I have been talking about procedure, or method. Now methods that are not used as a matter of habit, without any thought about the reasons behind them, are often tied to metaphysical beliefs. For example, a radical form of empiricism assumes either that humans are the measure of things or that they are in harmony with them. Applied consistently methodological rules may produce results which agree with the corresponding metaphysics. Luria's procedure is an example. It did not fail; it helped to build a subject which today is at the forefront of

---

11. 'Problems of Cosmogony and Cosmology', in F. Bertola, J.W. Sulentic and D.F. Madore (eds), *New Ideas in Astronomy*, Cambridge, 1988, p. 229.

12. N.R. Campbell, *Foundations of Science*, New York, 1957, p. 21.

research. Einstein's approach did not end in disaster; it led to one of the
most fascinating modern theories – general relativity. But methods are
not restricted to the area where they scored their first triumphs. Luria's
requirements, for example, also turned up in cosmology; they had
been used by Heber Curtis, in his 'grand debate' with Harlow Shapely;
by Ambarzumjan, who opposed empiricism to abstract principles; and
they are now being applied by Halton Arp, Margaret Geller and their
collaborators. Whatever the results, a world built up in the manner of
Luria has little in common with the world of Einstein and this world
again differs considerably from the world of Bohr. Johann Theodore Merz
describes in detail how abstract world-views using corresponding methods
produced results which slowly filled them with empirical content.[13] He
discusses the following views. First, *the astronomical view*, which rested
on mathematical refinements of action-at-a-distance laws and was
extended (by Coulomb, Neumann, Ampère and others) to electricity
and magnetism. Laplace's theory of capillarity was an outstanding
achievement of this approach. Secondly, *the atomic view*, which played
an important role in chemical research (example: stereochemistry) but
was also opposed by chemists. Thirdly, *the kinetic and mechanical view*,
which employed atoms in the area of heat and electric phenomena. For
some scientists atomism was the foundation of everything. Fourthly, *the
physical view*, which tried to achieve universality in a different way, on the
basis of general notions such as the notion of energy. It could be connected
with the kinetic view, but often was not. Physicians, physiologists and
chemists like Mayer, Helmholtz, du Bois Reymond and, in the practical
area, Liebig were outstanding representatives of this view in the second
half of the 19th century while Ostwald, Mach and Duhem extended it
into the 20th. Starting his description of *the morphological view*, Merz
writes:

> The different aspects of nature which I have reviewed in the foregoing
> chapters and the various sciences which have been elaborated by their
> aid, comprise what may appropriately be termed the abstract study of
> natural objects and phenomena. Though all the methods of reasoning
> with which we have so far become acquainted originated primarily
> through observation and the reflection over things natural, they have this

---

13. *A History of European Thought in the 19th Century* (first published 1904–12).

in common that they – for the purpose of examination – remove their objects out of the position and surroundings which nature has assigned to them: that they *abstract* them. This process of abstraction is either literally a process of removal from one place to another, from the great work – and storehouse of nature herself to the small workroom, the laboratory of the experimenter; or – where such removal is not possible – the process is carried out merely in the realm of contemplation; one or two special properties are noted and described, whilst the number of collateral data are for the moment disregarded. [A third method, not developed at the time, is the creation of 'unnatural' conditions and, thereby, the production of 'unnatural' phenomena.]

. . . There is, moreover, in addition to the aspect of convenience, one very powerful inducement for scientific workers to persevere in their process of abstraction. . . . This is the practical usefulness of such researches in the arts and industries. . . . The wants and creations of artificial life have thus proved the greatest incentives to the abstract and artificial treatment of natural objects and processes for which the chemical and electrical laboratories with the calculating room of the mathematician on the one side and the workshop and factory of the other, have in the course of the century become so renowned. . . .

There is, however, in the human mind an opposite interest which fortunately counteracts to a considerable extent the one-sided working of the spirit of abstraction in science. . . . This is the genuine love of nature, the consciousness that we lose all power if, to any great extent, we sever or weaken that connection which ties us to the world as it is – to things real and natural: it finds its expression in the ancient legend of the mighty giant who derived all his strength from his mother earth and collapsed if severed from her. . . . In the study of natural objects we meet [therefore] with a class of students who are attracted by things as they. . . . [Their] sciences are the truly descriptive sciences, in opposition to the abstract ones.[14]

I have quoted this description at length for it shows how different procedures rest on, and provide evidence for, different world-views. Finally, Merz mentions *the genetic view, the psychophysical view, the vitalistic view, the statistical view* together with their procedures and their findings.

---

14. Ibid., Vol. 2, New York, 1965, pp. 200f.

What can a single comprehensive 'world-view of science' or a single comprehensive idea of *science* offer under such circumstances?

It can offer a survey, a list similar to the list given by Merz, enumerating the achievements and drawbacks of the various approaches as well as the clashes between them and it can identify science with this complex and somewhat scattered wars on many fronts. Alternatively it can put one view on top and subordinate the others to it, either by pseudo-derivations, or by declaring them to be meaningless. Reductionists love to play that game. Or it can disregard the differences and present a paste job where each particular view and the results it has achieved is smoothly connected with the rest thus producing an impressive and coherent edifice – 'the' scientific world-view.

Expressing it differently we may say that the assumption of a single coherent world-view that underlies all of science is either a metaphysical hypothesis trying to anticipate a future unity, or a pedagogical fake; or it is an attempt to show, by a judicious up- and downgrading of disciplines, that a synthesis has already been achieved. This is how fans of uniformity proceeded in the past (see Plato's list of subjects in Chapter VII of his *Republic*), these are the ways that are still being used today. A more realistic account, however, would point out that '[t]here is no simple "scientific" map of reality – or if there were, it would be much too complicated and unwieldy to be grasped or used by anyone. But there are many different maps of reality, from a variety of scientific viewpoints'.[15]

It may be objected that we live in the 20th century, not in the 19th, and that many unifications which seemed impossible then have been achieved by now. Examples are statistical thermodynamics, molecular biology, quantum chemistry and superstrings. These are indeed flourishing subjects, but they have not produced the unity the phrase 'the' scientific view of the world insinuates. Actually, the situation is not very different from that which Merz had noticed in the 19th century. Truesdell and others continue the physical approach: Prandtl maligned Euler, Truesdell praises him for having provided rigorous concepts for research. Morphology, though given a low status by some and declared to be dead by others, has been revived by ecologists and by Lorenz's

---

15. John Ziman, *Teaching and Learning About Science and Society*, Cambridge, 1980, p. 19.

study of animal behaviour (which added *forms of motion* to the older *static forms*) and it has always been of importance in galactic research (Hubble's classification). Having been in the doghouse, cosmology is now being courted by high energy physicists but clashes with the philosophy of complementarity accepted by the same group. Commenting on the problem M. Kafatos and R. Nadeu write:

> The essential requirement of the Copenhagen interpretation that the experimental setup must be taken into account when making observations is seldom met in observations with cosmological import [though such observations rely on light, the paradigm case of complementarity].[16]

Moreover, the observations of Arp, M. Geller and others have thrown considerable doubt on the homogeneity assumption which plays a central role in it. Extended to 1,000 megaparsec, Geller's research may blow up the entire subject. We have a rabid materialism in some parts (molecular biology, for example), a modest to radical subjectivism in others (some versions of quantum measurement, anthropic principle). There are many fascinating results, speculations, attempts at interpretation and it is certainly worth knowing them. But pasting them together into a single coherent 'scientific' world-view, a procedure which has the blessings even of the Pope[17] – this is going too far. After all, who can say that the world which so strenuously resists unification really is as educators and metaphysicians want it to be – tidy, uniform, the same everywhere? Besides, as was shown in Chapters 3ff, a paste job eliminates precisely those conflicts that kept science going in the past and will continue inspiring its practitioners if preserved.

At this point some defenders of uniformity rise to a higher level. Science may be complex, they say, but it is still 'rational'. Now the word 'rational' can either be used as a collecting bag for a variety of procedures – this would be its nominalist interpretation – or it describes a general feature found in every single scientific action. I accept the first definition,

---

16. 'Complementarity and Cosmology', in M. Kafatos (ed.), *Bell's Theorem, Quantum Theory and the Conceptions of the Universe*, Dordrecht, 1980, p. 263.

17. See his message on the occasion of the 300th anniversary of Newton's *Principia*, published in *John Paul II on Science and Religion*, Notre Dame, 1990, esp. M6ff.

but I reject the second. In the second case rationality is either defined in a narrow way that excludes, say, the arts; then it also excludes large sections of the sciences. Or it is defined in a way that lets all of science survive; then it also applies to love-making, comedy and dogfights. There is no way of delimiting 'science' by something stronger and more coherent than a list.

I come to the second question – What's so great about science? There are various measures of greatness. *Popularity*, i.e. familiarity with some results and the belief that they are important, is one of them. Now it is true that despite periodic swings towards the sciences and away from them they are still in high repute with the general public – or, rather, not the scien*ces*, but a mythical monster 'science' (in the singular – in German it sounds even more impressive: *Die Wissenschaft*). For what the general public seems to assume is that the achievements they read about in the educational pages of their newspapers and the threats they seem to perceive come from a single source and are produced by a uniform procedure. They know that biology is different from physics which is different from geology. But these disciplines, it is assumed, arise when 'the scientific way' is applied to different topics; the scientific way itself, however, remains the same. I have tried to argue that scientific practice is much more diverse. Adding that scientists keep complaining about the scientific illiteracy of the general public and that by the 'general public' they mean the Western middle class, not Bolivian peasants (for example), we have to conclude that the popularity of science is a very doubtful matter indeed.

What about *practical advantages*? The answer is that 'science' sometimes works and sometimes doesn't. Some sciences (economic theory, for example) are in a pretty sorry shape. Others are sufficiently mobile to turn disaster into triumph. *The can do so because they are not tied to any particular method or world-view.* The fact that an approach is 'scientific' according to some clearly formulated criterion therefore is no guarantee that it will succeed. *Each case must be judged separately*, especially today, when the fear of industrial espionage, the wish to overtake competitors on the way to a Nobel Prize, the uneven distribution of funds, national rivalries, fear of accusations (of malpractice, plagiarism, waste of funds, etc.) put restrictions on what some dreamers, many philosophers among them, still regard as a 'free intellectual adventure'.[18] The question of *truth*, finally, remains unresolved. Love of truth is one of the strongest motives for replacing what really happens by a streamlined account, or, to express it in a less polite manner, love of truth is one of the strongest motives for lying to oneself and to others. Besides, the quantum theory seems to show, in the precise manner so much beloved by the admirers of science, that reality is either one, which means there are no observers and no things observed, or it is many, in which case what is found does not exist in itself but depends on the approach chosen.

What are the views that are being compared with science when it is declared to be superior? E.O. Wilson, the 'father' of sociobiology, writes:

> religion . . . will endure for a long time as a vital force in society. Like the mythical giant Antaeus who drew energy from his mother, the earth, religion cannot be defeated by those who may cast it down. The spiritual weakness of scientific naturalism is due to the fact that it has no such primal source of power. . . . So the time has come to ask: does a way exist to divert the power of religion into the services of the great new enterprise?[19]

For Wilson the main feature of the alternatives is that they have *power*. I regard this as a somewhat narrow characterization. World-views also answer questions about origins and purposes which sooner or later arise in almost every human being. Answers to these questions were available to Kepler and Newton and were used by them in their research; they are no longer available today, at least not within the sciences. They are part of non-scientific world-views which therefore have much to offer, also to scientists. When Western Civilization invaded what is now called the Third World it imposed its own ideas of a proper environment and a rewarding life. It thereby disrupted delicate patterns of adaptation and created problems that had not existed before. Both human decency and some appreciation of the many ways in which humans can live with nature prompted agents of development and public health to think in

---

18. This was realized by government advisers after the postwar euphoria had worn off. See Joseph Ben-David, *Scientific Growth*, Berkeley, 1991, p. 525, quoted above.

19. *On Human Nature*, Cambridge, Mass., 1972, pp. 192f.

more complex or, as some would say, more 'relativistic' ways. There exist approaches, the approach called 'Primary environmental Care' among them, which offer legal, political and scientific information but modified in accordance with the needs, the wishes and, what is most important, *the skills and the knowledge* of local populations.[20] Similarly, the movement called liberation theology has modified Church doctrine to bring it closer to the spiritual needs of the poor and disadvantaged, especially in South America.

Let me point out, incidentally, that not all ideas which seem repulsive to the prophets of a New Age come from science. The idea of a world machine and the related idea that nature is material to be shaped by man should not be blamed on modern, i.e. post-Cartesian, science. It is older and stronger than a purely philosophical doctrine could ever be. The expression 'world machine' is found in Pseudo Dionysius Areopagita, a mystic of unknown identity who wrote about 500 AD and had tremendous influence. Oresme, who died in 1382 as bishop of Lisieux, compares the universe to a vast mechanical clock set running by God so that 'all the wheels move as harmoniously as possible'. The sentiment can be easily understood: this was the time when mechanical clocks 'of astounding intricacy and elaboration' were constructed all over Europe – every town was supposed to have one. Lynn White Jr., from whose book I have taken this information, also describes the change of attitude that occurred in the Carolingian Age:

> The old Roman Calendars had occasionally shown genre scenes of human activity but the dominant tradition (which continued in Byzantium) was to depict the months as passive personifications being symbols of attributes. The new Carolingian calendar which set the pattern for the Middle Ages . . . shows a coercive attitude towards natural resources. . . . The pictures [are about] scenes of ploughing, harvesting, woodchopping; people knocking down acorns for the pigs, pig slaughtering. Man and Nature are now two things and man is the master.[21]

---

20. *Lessons Learned in Community-Based Environmental Management*, Proceedings of the 1990 Primary Environmental Care Workshop, ed. Grazia Borrini, International Course for Primary Health Care Managers at District Level in Developing Countries, Istituto Superiore di Sanità, Rome, 1991. For a more popular presentation see Grazia Borrini, 'Primary Environmental Care: For Environmental Advocates and Policy-Makers', *UNESCO Courier*, forthcoming.

21. *Mediaeval Technology and Social Change*, Oxford, 1960, pp. 56f.

To sum up: there is no 'scientific world-view', just as there is no uniform enterprise 'science' – except in the minds of metaphysicians, schoolmasters and politicians trying to make their nation competitive. Still, there are many things we can learn from the sciences. But we can also learn from the humanities, from religion and from the remnants of ancient traditions that survived the onslaught of Western Civilization. No area is unified and perfect, few areas are repulsive and completely without merit. There is no objective principle that could direct us away from the supermarket 'religion' or the supermarket 'art' towards the more modern, and much more expensive supermarket 'science'. Besides, there are large areas of knowledge and action in which we use procedures without any idea as to their comparative excellence. An example is medicine, which, though not a science, has increasingly been connected with scientific research. There are many fashions and schools in medicine just as there are many fashions and schools in psychology. It follows, first, that the idea of a comparison of 'Western medicine' with other medical procedures does not make sense. Secondly, such a comparison is often against the law, even if there should be volunteers: a test is legally impossible. Adding to this that health and sickness are culture-dependent concepts, we see that there are domains, such as medicine with no scientific answer to question 2. This is not really a drawback. The search for objective guidance is in conflict with the idea of individual responsibility which allegedly is an important ingredient of a 'rational' or scientific age. It shows fear, indecision, a yearning for authority and a disregard for the new opportunities that now exist: we can build world-views on the basis of a personal choice and thus unite, for ourselves and for our friends, what was once separated by a series of historical accidents.[22]

On the other hand, we can agree that in a world full of scientific products scientists may be given a special status just as henchmen had a special status at times of social disorder or priests had when being a citizen coincided with being the member of a single universal Church.

---

22. Wolfgang Pauli, who was deeply concerned about the intellectual situation of the time, demanded that science and religion again be united: letter to M. Fierz, 8 August 1948. I agree but would add, entirely in the spirit of Pauli, that the unification should be a personal matter; it should not be prepared by philosophical-scientific alchemists of the mind and imposed by their minions in education. (It is different in the Third World where a strong faith still survives.)

We can also agree that appealing to a chimaera (such as that of a uniform and coherent 'scientific world-view') can have important political consequences. In 1854 Commander Perry, using force, opened the ports of Hakodate and Shimoda to American ships for supply and trade. This event demonstrated the military inferiority of Japan. The members of the Japanese enlightenment of the early 1870s, Fukuzawa among them, now reasoned as follows. Japan can keep its independence only if it becomes stronger. It can become stronger only with the help of science. It will use science effectively only if it does not just practise science but also believes in the underlying ideology. To many traditional Japanese this ideology – 'the' scientific world-view – was barbaric. But, so the followers of Fukuzawa argued, it was necessary to adopt barbaric ways, to regard them as advanced, to introduce the whole of Western Civilization in order to survive.[23] Having been thus prepared Japanese scientists soon branched out as their Western colleagues had done before and falsified the uniform ideology that had started the development. The lesson I draw from this sequence of events is that a uniform 'scientific view of the world' may be useful *for people doing science* – it gives them motivation without tying them down. It is like a flag. Though presenting a single pattern, it makes people do many different things. However, *it is a disaster for outsiders* (philosophers, fly-by-night mystics, prophets of a New Age). It suggests to them the most narrowminded religious commitment and encourages a similar narrowmindedness on their part.

What I have said so far already contains my answer to question 3: a community *will* use science and scientists in a way that agrees with its values and aims and it will correct the scientific institutions in its midst to bring them closer to these aims. The objection that science is self-correcting and thus needs no outside interference overlooks, first, that every enterprise is self-correcting (look at what happened to the Catholic Church after Vatican II) and, secondly, that in a democracy the self-correction of the whole which tries to achieve more humane ways of living overrules the self-correction of the parts which has a more narrow aim – unless the parts are given temporary independence. Hence in a democracy local populations not only will, but also *should*, use the

---

23. Details in Carmen Blacker, *The Japanese Enlightenment*, Cambridge, 1969. For the political background see Chapters 3 and 4 of Richard Storry, *A History of Modern Japan*, Harmondsworth, 1982.

sciences in ways most suitable to them. The objection that citizens do not have the expertise to judge scientific matters overlooks that important problems often lie across the boundaries of various sciences so that scientists within these sciences don't have the needed expertise either. Moreover, doubtful cases always produce experts for the one side, experts for the other side, and experts in between. But the competence of the general public could be vastly improved by an education that exposes expert fallibility instead of acting as if it did not exist.

The point of view underlying this book is not the result of a well-planned train of thought but of arguments prompted by accidental encounters. Anger at the wanton destruction of cultural achievements from which we all could have learned, at the conceited assurance with which some intellectuals interfere with the lives of people, and contempt for the treacly phrases they use to embellish their misdeeds was, and still is the motive force behind my work.

The problem of knowledge and education in a free society first struck me during my tenure of a state fellowship at the Weimar Institut zur Methodologischen Erneuerung des Deutschen Theaters (1946), which was a continuation of the Deutsches Theater Moskau under the directorship of Maxim Vallentin. Staff and students of the Institut periodically visited theatres in Eastern Germany.[1] A special train brought us from city to city. We arrived, dined, talked to the actors, watched two or three plays. After each performance the public was asked to remain seated while we started a discussion of what we had just seen. There

---

1. Like many people of my generation I was involved in the Second World War. This event had little influence on my thinking. For me the war was a nuisance, not a moral problem. Before the war I had intended to study astronomy, acting and singing and to practise these professions simultaneously. I had excellent teachers (Adolf Vogel, my singing teacher, had an international reputation and taught outstanding opera singers such as Norman Bayley) and had just overcome some major vocal difficulties when I received my draft notice (I was eighteen at the time). How inconvenient, I thought. Why the hell should I participate in the war games of a bunch of idiots? How do I get out of it? Various attempts misfired and I became a soldier. I applied for officers' training to avoid bullets as long as possible. The attempt was not entirely successful; I was a lieutenant before the war had come to an end and found myself in the middle of the German retreat in Poland and then in East Germany, surrounded by fleeing civilians, infantry units, tanks, Polish auxiliaries whom I suddenly commanded (the higher officers quickly disappeared when matters became sticky). The whole colourful chaos then appeared to me like a stage and I became careless. A bullet hit me on my right hand, a second bullet grazed my face, a third got stuck in my spine, I fell to the ground, unable to rise, but with the happy thought 'the war is over for me, now at last I can return to singing and my beloved astronomy books'. It was only much later that I became aware of the moral problems of the entire age. It seems to me that these problems are still with us. They arise whenever an individual or a group objectivizes its own personal conceptions of a good life and acts accordingly. See *Farewell to Reason*, pp. 309ff. This explains the occasional violence of my arguments.

were classical plays, but there were also new plays which tried to analyse recent events. Most of the time they dealt with the work of the resistance in Nazi Germany. They were indistinguishable from earlier Nazi plays eulogizing the activity of the Nazi underground in democratic countries. In both cases there were ideological speeches, outbursts of sincerity and dangerous situations in the cops and robbers tradition. This puzzled me and I commented on it in the debates: how should a play be structured so that one recognizes it as presenting the 'good side'? What has to be added to the action to make the struggle of the resistance fighter appear morally superior to the struggle of an illegal Nazi in Austria before 1938? It is not sufficient to give him the 'right slogans' for then we take his superiority for granted, we do not show wherein it consists. Nor can his nobility, his 'humanity' be the distinguishing mark; every movement has scoundrels as well as noble people among its followers. A playwright may of course decide that sophistication is luxury in moral battles and give a black–white account. He may lead his followers to victory but at the expense of turning them into barbarians. What, then, is the solution? At the time I opted for Eisenstein and ruthless propaganda for the 'right cause'. I don't know whether this was because of any deep conviction of mine, or because I was carried along by events, or because of the magnificent art of Eisenstein. Today I would say that the choice must be left to the audience. The playwright presents characters and tells a story. If he errs it should be on the side of sympathy for his scoundrels, for circumstances and suffering play as large a role in the creation of evil and evil intentions as do those intentions themselves, and the general tendency is to emphasize the latter. The playwright (and his colleague, the teacher) must not try to anticipate the decision of the audience (of the pupils) or replace it by a decision of his own if they should turn out to be incapable of making up their own minds. *Under no circumstances must he try to be a 'moral force'.* A moral force, whether for good or for evil, turns people into slaves and slavery, even slavery in the service of The Good, or of God Himself, is the most abject condition of all. This is how I see the situation today. However, it took me a long time before I arrived at this view.

After a year in Weimar I wanted to add the sciences and the humanities to the arts, and the theatre. I left Weimar and became a student (history, auxiliary sciences) at the famous Institut fur Osterreichische Geschichtforschung which is part of the University of Vienna. Later on I added physics and astronomy and so finally returned to the subjects

I had decided to pursue before the interruptions of the Second World War.

There were the following 'influences'.

(1) The *Kraft Circle*. Many of us science and engineering students were interested in the foundations of science and in broader philosophical problems. We visited philosophy lectures. The lectures bored us and we were soon thrown out because we asked questions and made sarcastic remarks. I still remember Professor Heintel advising me with raised arms: 'Herr Feyerabend, entweder sie halten das Maul, oder sie verlassen den Vorlesungsaal!' We did not give up and founded a philosophy club of our own. Victor Kraft, one of my teachers, became our chairman. The members of the club were mostly students,[2] but there were also visits by faculty members and foreign dignitaries. Juhos, Heintel, Hollitscher, von Wright, Anscombe, Wittgenstein came to our meetings and debated with us. Wittgenstein, who took a long time to make up his mind and then appeared over an hour late, gave a spirited performance and seemed to prefer our disrespectful attitude to the fawning admiration he encountered elsewhere. Our discussions started in 1949 and proceeded with interruptions up to 1952 (or 1953). Almost the whole of my thesis was presented and analysed at the meetings and some of my early papers are a direct outcome of these debates.

(2) The Kraft Circle was part of an organization called the *Austrian College Society*. The Society had been founded in 1945 by Austrian resistance fighters[3] to provide a forum for the exchange of scholars and ideas and so to prepare the political unification of Europe. There were seminars, like the Kraft Circle, during the academic year and international meetings during the summer. The meetings took place (and still take place) in Alpbach, a small mountain village in Tirol. Here I met outstanding scholars, artists, politicians and I owe my academic career to the friendly help of some of them. I also began suspecting that what

---

2. Many of them have now become scientists or engineers. Johnny Sagan is Professor of Mathematics at the University of Illinois, Henrich Eichhorn director of New Haven observatory, Goldberger de Buda adviser to electronic firms, while Erich Jantsch, who died much too soon, met members of our circle at the astronomical observatory and later became a guru of dissident or pseudo-dissident scientists, trying to use old traditions for new purposes.

3. Otto Molden, brother of Fritz Molden of the Molden publishing house, was for many years the dynamic leader and organizer.

counts in a public debate are not arguments but certain ways of presenting one's case. To test the suspicion I intervened in the debates, defending absurd views with great assurance. I was consumed by fear – after all, I was just a student surrounded by bigshots – but having once attended an acting school I proved the case to my satisfaction. The difficulties of *scientific* rationality were made very clear by

(3) *Felix Ehrenhaft*, who arrived in Vienna in 1947. We, the students of physics, mathematics, astronomy, had heard a lot about him. We knew that he was an excellent experimenter and that his lectures were performances on a grand scale which his assistants had to prepare for hours in advance. We knew that he had taught theoretical physics, which was as exceptional for an experimentalist then as it is now. We were also familiar with the persistent rumours that denounced him as a charlatan. Regarding ourselves as defenders of the purity of physics we looked forward to exposing him in public. At any rate our curiosity was aroused – and we were not disappointed.

Ehrenhaft was a mountain of a man, full of vitality and unusual ideas. His lectures compared favourably (or unfavourably, depending on the point of view) with the more refined performances of his colleagues. 'Are you dumb? Are you stupid? Do you really agree with everything I say?' he shouted at us who had intended to expose him but sat in silent astonishment at his performance. The question was more than justified for there were large chunks to swallow. Relativity and quantum theory were rejected at once, and almost as a matter of course, for being idle speculation. In this respect Ehrenhaft's attitude was very close to that of Stark and Lenard, both of whom he mentioned with approval. But he went further and criticized the foundations of classical physics as well. The first thing to be removed was the law of inertia: undisturbed objects instead of going in a straight line were supposed to move in a helix. Then came a sustained attack on the principles of electromagnetic theory and especially on the equation div B = 0. Many years before the fundamental debate he produced convincing evidence for mesoscopic magnetic monopoles. Then new and surprising properties of light were demonstrated – and so on and so forth. Each demonstration was accompanied by a few gently ironical remarks on 'school physics' and the 'theoreticians' who built castles in the air without considering the experiments which Ehrenhaft devised and continued devising in all fields and which produced a plethora of inexplicable results.

We had soon an opportunity to witness the attitude of orthodox physicists. In 1949 Ehrenhaft came to Alpbach. In that year Popper conducted a seminar on philosophy, Rosenfeld and M.H.L. Pryce taught physics and philosophy of physics (mainly from Bohr's comments on Einstein which had just appeared), Max Hartmann biology, Duncan Sandys talked on problems of British politics, Hayek on economics and so on. There was Hans Thirring, the senior theoretical physicist from Vienna, a superb teacher who constantly tried to impress on us that there were more important things than science, who had taught physics to Feigl, Popper as well as the present author and was an early and very active member of the peace movement. His son Walter Thirring, now Professor of Theoretical Physics in Vienna, was also present – a very distinguished audience and a very critical one.

Ehrenhaft came well prepared. He set up a few of his simple experiments in one of the country houses of Alpbach and invited everyone he could lay hands on to have a look. Every day from two or three in the afternoon participants went by in an attitude of wonder and left the building (if they were theoretical physicists, that is) as if they had seen something obscene. Apart from these physical preparations Ehrenhaft also carried out, as was his habit, a beautiful piece of advertising. The day before his lecture he attended a fairly technical talk by von Hayek on 'The Sensory Order' (now available, in expanded form, as a book). During the discussion he rose, bewilderment and respect in his face, and started in a most innocent voice: 'Dear Professor Hayek. This was a marvellous, an admirable, a most learned lecture. I did not understand a single world. . . .' Next day his lecture had an overflow audience.

In this lecture Ehrenhaft gave a brief account of his discoveries, adding general observations on the state of physics. 'Now, gentlemen,' he concluded triumphantly, turning to Rosenfeld and Pryce who sat in the front row, 'what can you say?' And he answered immediately. 'There is nothing at all you can say with all your fine theories. Sitzen müssen sie bleiben! Still müssen sie sein!'

The discussion, as was to be expected, was quite turbulent and it was continued for days with Thirring and Popper taking Ehrenhaft's side against Rosenfeld and Pryce. Confronted with the experiments the latter occasionally acted as some of Galileo's opponents must have acted when confronted with the telescope. They pointed out that no conclusions could be drawn from complex phenomena and that a detailed analysis was needed. In short, the phenomena were a *Dreckeffect* – a word that was heard quite frequently in the arguments. What was our attitude in the face of all this commotion?

None of us was prepared to give up theory or to deny its excellence. We founded a Club for the Salvation of Theoretical Physics and started discussing simple experiments. It turned out that the relation between theory and experiment was much more complex than is shown in textbooks and even in research papers. There are a few paradigmatic cases where the theory can be applied without major adjustments but the rest must be dealt with by occasionally rather doubtful approximations and auxiliary assumptions.[4] I find it interesting to remember how little effect all this had on us at the time. We continued to prefer abstractions, as if the difficulties we had found had not been an expression of the nature of things but could be removed by some ingenious device, yet to be discovered. Only much later did Ehrenhaft's lesson sink in and our attitude at the time as well as the attitude of the entire profession provided me then with an excellent illustration of the nature of scientific rationality.

(4) *Philipp Frank* came to Alpbach a few years after Ehrenhaft. He undermined common ideas of rationality in a different way by showing that the arguments against Copernicus had been perfectly sound and in agreement with experience while Gailieo's procedures were 'unscientific' when viewed from a modern standpoint. His observations fascinated me and I examined the matter further. Chapters 8 to 11 are a late result of this study (I am a slow worker). Frank's work has been treated quite unfairly by philosophers like Putnam who prefer simplistic models to the analysis of complex historical events. Also his ideas are now commonplace. But it was he who announced them when almost everyone thought differently.

(5) In Vienna I became acquainted with some of the foremost Marxist intellectuals. This was the result of an ingenious PR job by Marxist students. They turned up – as did we – at all major discussions whether the subject was science, religion, politics, the theatre, or free love. They talked to those of us who used science to ridicule the rest – which was then my favourite occupation – invited us to discussions of their own and introduced us to Marxist thinkers from all fields. I came to know Berthold Viertel, the director of the Burgtheater, Hanns Eisler, the composer and

---

4. See Chapter 5 on ad hoc approximations.

music theoretician, and *Walter Hollitscher*, who became a teacher and, later on, one of my best friends. When starting to discuss with Hollitscher I was a raving positivist, I favoured strict rules of research and had only a pitying smile for the three basic principles of dialectics which I had read in Stalin's little pamphlet on dialectical and historical materialism. I was interested in the realist position, I had tried to read every book on realism I could lay hands on (including Külpe's excellent *Realisierung* and, of course, *Materialism and Empiriocriticism*) but I found that the arguments for realism worked only when the realist assumption had already been introduced. Külpe, for example, emphasized the distinction between impression and the thing the impression is about. The distinction gives us realism only if it characterizes real features of the world – which is the point at issue. Nor was I convinced by the remark that science is an essentially realistic enterprise. Why should science be chosen as an authority? And were there not positivistic interpretations of science? The so-called 'paradoxes' of positivism, however, which Lenin exposed with such consummate skill, did not impress me at all. They arose only if the positivist and the realist mode of speech were mixed and they exposed their difference. They did not show that realism was better, though the fact that realism came with common sense gave the impression that it was.

Hollitscher never presented an argument that would lead, step by step, from positivism into realism and he would have regarded the attempt to produce such an argument as philosophical folly. He rather developed the realist position itself, illustrated it by examples from science and common sense, showed how closely it was connected with scientific research and everyday action and so revealed its strength. It was of course always possible to turn a realistic procedure into a positivistic procedure by a judicious use of *ad hoc* hypotheses and *ad hoc* meaning changes and I did this frequently, and without shame (in the Kraft Circle we had developed such evasions into a fine art). Hollitscher did not raise semantic points, or points of method, as a critical rationalist might have done, he continued to discuss concrete cases until I felt rather foolish with my abstract objections. For I saw now how closely realism was connected with facts, procedures, principles I valued and that it *had helped to bring them about* while positivism merely *described* the results in a rather complicated way after they had been found: realism had fruits, positivism had none. This at least is how I would speak today, long *after* my realist conversion. At the time I became a realist not because I was convinced by any particular argument, but because the sum total of realism plus the arguments in favour of it plus the ease with which it could be applied to science and many other things I vaguely felt but could not lay a finger on[5] finally *looked better to me* than the sum total of positivism plus the arguments one could offer for it plus . . . etc., etc. The comparison and the final decision had much in common with the comparison of life in different countries (weather, character of people, melodiousness of language, food, laws, institutions, weather, etc., etc.) and the final decision to take a job and to start life in one of them. Experiences such as these have played a decisive role in my attitude towards rationalism.

While I accepted realism I did not accept dialectics and historical materialism – my predilection for abstract arguments (another positivist hangover) was still too strong for that. Today Stalin's rules seem to me preferable by far to the complicated and epicycle-ridden standards of our modern friends of reason.

From the very beginning of our discussion Hollitscher made it clear that he was a communist and that he would try to convince me of the intellectual and social advantages of dialectical and historical materialism. There was none of the mealy-mouthed 'I may be wrong, you may be right – but together we shall find the truth' talk with which 'critical' rationalists embroider their attempts at indoctrination but which they forget the moment their position is seriously endangered. Nor did Hollitscher use unfair emotional or intellectual pressures. Of course, he criticized my attitude but our personal relations have not suffered from my reluctance to follow him in every respect. This is why Walter Hollitscher is a teacher while Popper, whom I also came to know quite well, is a mere propagandist.

At some point of our acquaintance Hollitscher asked me whether I would like to become a production assistant of Brecht – apparently there

---

5. I remember that Reichenbach's answer to Dingler's account of relativity played an important part: Dingler extrapolated from what could be achieved by simple mechanical operations (manufacture of a Euclidean plain surface, for example) while Reichenbach pointed out how the actual structure of the world would modify the results of these operations in the large. It is of course true that Reichenbach's account can be interpreted as a more efficient predictive machine and that it seemed impressive to me only because I did not slide into such an interpretation. Which shows to what extent the force of arguments depends on irrational changes of attitude.

was a position available and I was being considered for it. I declined. For a while I thought that this was one of the biggest mistakes of my life. Enriching and changing knowledge, emotions, attitudes through the arts now seems to me a much more fruitful enterprise and also much more humane than the attempt to influence minds (and nothing else) by words (and nothing else). Reading about the tensions inside the Brecht Circle, the almost religious attitude of some of its members, I now think that I escaped just in time.

(6) During a lecture (on Descartes) I gave at the Austrian College Society I met *Elizabeth Anscombe*, a powerful and, to some people, forbidding British philosopher who had come to Vienna to learn German for her translation of Wittgenstein's works. She gave me manuscripts of Wittgenstein's later writings and discussed them with me. The discussions extended over months and occasionally proceeded from morning over lunch until late into the evening. They had a profound influence upon me though it is not at all easy to specify particulars. On one occasion which I remember vividly Anscombe, by a series of skilful questions, made me see how our conception (and even our perceptions) of well-defined and apparently self-contained facts may depend on circumstances not apparent in them. There are entities such as physical objects which obey a 'conservation principle' in the sense that they retain their identity through a variety of manifestations and even when they are not present at all, while other entities such as pains and after-images are 'annihilated' with their disappearance. The conservation principles may change from one developmental stage of the human organism to another[6] and they may be different for different languages (see Whorf's 'covert classifications' as described in Chapter 16). I conjectured that such principles would play an important role in science, that they might change during revolutions and that deductive relations between pre-revolutionary and post-revolutionary theories might be broken off as a result. I explained this early version of incommensurability in Popper's seminar (1952) and to a small group of people in Anscombe's flat in Oxford (also in 1952 with Geach, von Wright and L.L. Hart present) but I was not able to arouse much enthusiasm on either occasion. Wittgenstein's emphasis on the need for concrete research and his objections to abstract reasoning ('Look, don't think!') somewhat clashed with my own tendency towards

---

6. See Chapter 16, text to footnotes 12ff.

abstractness, and the papers in which his influence is noticeable are therefore mixtures of concrete examples and sweeping principles.[7] Wittgenstein was prepared to take me on as a student in Cambridge but he died before I arrived. Popper became my supervisor instead.

(7) I had met *Popper* in Alpbach in 1948. I admired his freedom of manners, his cheek, his disrespectful attitude towards the German philosophers who gave the proceedings weight in more senses than one, his sense of humour (yes, the relatively unknown Karl Popper of 1948 was very different from the established Sir Karl of later years) and I also admired his ability to restate ponderous problems in simple and journalistic language. Here was a free mind, joyfully putting forth his ideas, unconcerned about the reaction of the 'professionals'. Things were different as regards these ideas themselves. The members of our circle knew deductivism from Kraft who had written about it before Popper,[8] and the falsificationist philosophy was taken for granted in the physics seminar of the conference under the chairmanship of Arthur March and so we did not understand what all the fuss was about. 'Philosophy must be in a desperate state', we said, 'if trivialities such as these can count as major discoveries.' Popper himself did not seem to think too much of his philosophy of science at the time for when asked to send us a list of publications he included the *Open Society* but not the *Logic of Scientific Discovery*.

While in London I read Wittgenstein's *Philosophical Investigations* in detail. Being of a rather pedantic turn of mind I rewrote the book so that it looked more like a treatise with a continuous argument. Part of this treatise was translated by Anscombe into English and published as a review by *Philosophical Review* in 1955. I also visited Popper's seminar at the LSE. Popper's ideas were similar to those of Wittgenstein but they were more abstract and anaemic. This did not deter me but increased my own tendencies to abstraction and dogmatism. At the end of my stay in London Popper invited me to become his assistant. I declined despite the fact that I was broke and did not know where my next meal was going to come from. My decision was not based on any clearly recognizable

---

7. For details see my comments on these papers in *Der Wissenschaftstheoretische Realismus und die Autorität der Wissenschaften*, Vieweg Wiesbaden, 1978.

8. See my review of Kraft's *Erkenntnislehre* in *BJPS*, Vol. 13, 1963, pp. 319ff., esp. p. 321, second paragraph. See also the references in Popper, *Logic of Scientific Discovery*. Mill's *System of Logic*, Vol. 2, London, 1879, Chapter 14, gives a detailed account of the procedure.

train of thought but I guess that, having no fixed philosophy, I preferred stumbling around in the world of ideas at my own speed to being guided by the ritual of a 'rational debate'. Again I was lucky. Joseph Agassi who got the job did not have much privacy. Two years later Popper, Schrödinger and my own big mouth got me a job in Bristol where I started lecturing on the philosophy of science.

(8) I had studied theatre, history, mathematics, physics and astronomy; I had never studied philosophy. The prospect of having to address a large audience of eager young people did not exactly fill my heart with joy. One week before the lectures started I sat down and wrote everything I knew on a piece of paper. It hardly filled a page. Agassi came up with some excellent advice: 'Look, Paul,' he said, 'the first line, this is your first lecture; the second line, this is your second lecture – and so on.' I took his advice and fared rather well except that my lectures became a stale collection of wisecracks from Wittgenstein, Bohr, Popper, Dingler, Eddington and others. While in Bristol I continued my studies of the quantum theory. I found that important physical principles rested on methodological assumptions that are violated whenever physics advances: physics gets authority from ideas it propagates but never obeys in actual research, methodologists play the role of publicity agents whom physicists hire to praise their results but whom they would not permit access to the enterprise itself. That falsificationism is not a solution became very clear in discussions with David Bohm who gave a Hegelian account of the relation between theories, their evidence, and their successors.[9] The material of Chapter 3 is the result of these discussions (I first published it in 1961).[10] Kuhn's remarks on the omnipresence of

---

9. I have explained the Hegelianism of Bohm in the essay 'Against Method' which appeared in Vol. 4 of the *Minnesota Studies for the Philosophy of Science*, 1970.

10. Popper once remarked (in a discussion at the Minnesota Center for the Philosophy of Science in the year 1962) that the example of Brownian motion is just another version of Duhem's example (conflict between specific laws such as Kepler's laws and general theories such as Newton's theory). But there is a most important difference. The deviations from Kepler's laws are in principle observable ('in principle' meaning 'given the known laws of nature'), while the microscopic deviations from the second law of thermodynamics are not (measuring instruments are subjected to the same fluctuations as the things they are supposed to measure). Here we *cannot* do without an alternative theory. See Chapter 4, footnote 2.

anomalies fitted these difficulties rather nicely[11] but I still tried to find general rules that would cover all cases[12] and non-scientific developments as well.[13] Two events made me realize the futility of such attempts. One was a discussion with Professor C.F. von Weizsäcker in Hamburg (1965) on the foundations of the quantum theory. Von Weizsäcker showed how quantum mechanics arose from concrete research while I complained, on general methodological grounds, that important alternatives had been omitted. The arguments supporting my complaint were quite good – they are the arguments summarized in Chapter 3 – but it was suddenly clear to me that, imposed without regard to circumstances, they were a hindrance rather than a help: a person trying to solve a problem whether in science or elsewhere *must be given complete freedom* and cannot be restricted by any demands, norms, however plausible they may seem to the logician or the philosopher who has thought them out in the privacy of his study. Norms and demands must be checked by research, not by appeal to theories of rationality. In a lengthy article[14] I explained how Bohr had used this philosophy and how it differs from more abstract procedures. Thus Professor von Weizsäcker has prime responsibility for my change to 'anarchism' – though he was not at all pleased when I told him so in 1977.

(9) The second event that prompted me to move away from rationalism and to become suspicious of all intellectual pretensions was quite different. To explain it, let me start with some general observations. The way in which social problems, problems of energy distribution, ecology, education, care for the old and so on are 'solved' in First World societies can be roughly described in the following way. A problem arises. Nothing is done about it. People get concerned. Politicians broadcast this concern. Experts are called in. They develop theories and plans based on them. Power-groups with experts of their own effect various modifications until a watered down version is accepted and realized. The role of experts in this process has gradually increased. We have now a situation where

---

11. I read Kuhn's book in manuscript in 1960 and discussed it extensively with Kuhn.

12. See the account in 'Reply to Criticism', *Boston Studies*, Vol. 2, 1965.

13. See 'On the Improvement of the Sciences and the Arts and the Possible Identity of the Two' in *Boston Studies*, Vol. 3, 1967.

14. 'On a Recent Critique of Complementarity', *Philosophy of Science* 1968/69 (two parts).

social and psychological *theories* of human thought and action have taken the place of this thought and action itself. Instead of asking the people involved in a problematic situation, developers, educators, technologists and sociologists get their information about 'what these people really want and need' from theoretical studies carried out by their esteemed colleagues in what they think are the relevant fields. Not live human beings, but abstract models are consulted; not the target population decides, but the producers of the models. Intellectuals all over the world take it for granted that their models will be more intelligent, make better suggestions, have a better grasp of the reality of humans than these humans themselves. What has this situation got to do with me?

From 1958 to 1990 I was a Professor of Philosophy at the University of California in Berkeley. My function was to carry out the educational policies of the State of California, which means I had to teach people what a small group of white intellectuals had decided was knowledge. I hardly ever thought about this function and I would not have taken it very seriously had I been informed. I told the students what I had learned, I arranged the material in a way that seemed plausible and interesting to me – and that was all I did. Of course, I had also some 'ideas of my own' – but these ideas moved in a fairly narrow domain (though some of my friends said even then that I was going batty).

In the years around 1964 Mexicans, blacks, Indians entered the university as a result of new educational policies. There they sat, partly curious, partly disdainful, partly simply confused, hoping to get an 'education'. What an opportunity for a prophet in search of a following! What an opportunity, my rationalist friends told me, to contribute to the spreading of reason and the improvement of mankind! What a marvellous opportunity for a new wave of enlightenment! I felt very differently. For it now dawned on me that the intricate arguments and the wonderful stories I had so far told to my more or less sophisticated audience might just be dreams, reflections of the conceit of a small group who had succeeded in enslaving everyone else with their ideas. Who was I to tell these people what and how to think? I did not know their problems, though I knew they had many. I was not familiar with their interests, their feelings, their fears, though I knew that they were eager to learn. Were the arid sophistications which philosophers had managed to accumulate over the ages and which liberals had surrounded with schmaltzy phrases to make them palatable the right thing to offer to people who had been robbed of their land, their culture, their dignity and who were now supposed first to absorb and then to repeat the anaemic ideas of the mouthpieces of their oh so human captors? They wanted to know, they wanted to learn, they wanted to understand the strange world around them – did they not deserve better nourishment? Their ancestors had developed cultures of their own, colourful languages, harmonious views of the relation between people, and between people and nature whose remnants are a living criticism of the tendencies of separation, analysis, self-centredness inherent in Western thought. These cultures have important achievements in what is today called sociology, psychology, medicine, they express ideals of life and possibilities of human existence. Yet *they were never examined with the respect they deserved* except by a small number of outsiders; they were ridiculed and replaced as a matter of course first by the religion of brotherly love and then by the religion of science or else they were defused by a variety of 'interpretations'. Now there was much talk of liberation, of racial equality – but what did it mean? Did it mean the equality of these traditions and the traditions of the white man? It did not. Equality meant that the members of different races and cultures now had the wonderful chance to participate in the white man's manias, they had the chance to participate in his science, his technology, his medicine, his politics. These were the thoughts that went through my head as I looked at my audience and they made me recoil in revulsion and terror from the task I was supposed to perform. For the task – this now became clear to me – was that of a very refined, very sophisticated slavedriver. And a slavedriver I did not want to be.

Experiences such as these convinced me that intellectual procedures which approach a problem through concepts are on the wrong track, and I became interested in the reasons for the tremendous power this error has now over minds. I started examining the rise of intellectualism in Ancient Greece and the causes that brought it about. I wanted to know what it is that makes people who have a rich and complex culture fall for dry abstractions and multilate their traditions, their thought, their language so that they can accommodate the abstractions. I wanted to know how intellectuals manage to get away with murder – for it is murder, murder of minds and cultures that is committed year in, year out at schools, universities, educational missions in foreign countries. The trend must be reversed, I thought: we must start learning from those we have enslaved, for they have much to offer and, at any rate, they have the right to live as they see fit even if they are not as pushy about their rights and their views as their Western conquerors have always been. In 1964–5, when these ideas first occurred to me, I tried to find an *intellectual* solution to my misgivings, that is, I took it for granted that it was up to *me* and the likes of me to devise educational policies for other people. I envisaged a new kind of education that would live from a rich reservoir of different points of view permitting the choice of traditions most advantageous to the individual. The teacher's task would consist in facilitating the choice, not in replacing it by some 'truth' of his own. Such a reservoir, I thought, would have much in common with a *theatre* of ideas as imagined by Piscator and Brecht and it would lead to the development of a great variety of means of presentation. The 'objective' scientific account would be one way of presenting a case, a play another way (remember that for Aristotle tragedy is 'more philosophical' than history because it reveals the *structure* of the historical process and not only its accidental details), a novel still another way. Why should knowledge be shown in the garment of academic prose and reasoning? Had not Plato observed that written sentences in a book are but transitory stages of a complex process of growth that contains gestures, jokes, asides, emotions and had he not tried to catch this process by means of the dialogue? And were there not different forms of knowledge, some much more detailed and realistic than what arose as 'rationalism' in the 7th and 6th century in Greece? Then there was *Dadaism*. I had studied Dadaism after the Second World War. What attracted me to this movement was the style its inventors used when not engaged in Dadaistic activities. It was clear, luminous, simple without being banal, precise without being narrow; it was a style adapted to the expression of thought as well as of emotion. I connected this style with the Dadaistic exercises themselves. Assume you tear language apart, you live for days and weeks in a world of cacophonic sounds, jumbled words, nonsensical events. Then, after this preparation, you sit down and write: 'the cat is on the mat'. This simple sentence which we usually utter without thought, like talking machines (and much of our talk is indeed routine), now seems like the creation of an entire world: God said let there be light, and there was light. Nobody in modern times has understood the miracle of language and thought as well as the Dadaists, for nobody has been able to imagine, let alone create, a world in which they play no role. Having discovered the nature of a *living order*, of a reason that is not merely mechanical, the

Dadaists soon noticed the deterioration of such an order into routine. They diagnosed the deterioration of language that preceded the First World War and created the mentality that made it possible. After the diagnosis their exercises assumed another, more sinister meaning. They revealed the frightening similarity between the language of the foremost commercial travellers in 'importance', the language of philosophers, politicians, theologians, and brute inarticulation. The praise of honour, patriotism, truth, rationality, honesty that fills our schools, pulpits, political meetings *imperceptibly merges into inarticulation* no matter how much it has been wrapped into literary language and no matter how hard its authors try to copy the style of the classics, and the authors themselves are in the end hardly distinguishable from a pack of grunting pigs. Is there a way to prevent such deterioration? I thought there was. I thought that regarding all achievements as transitory, restricted *and personal* and every truth as *created* by our love for it and not as 'found' would prevent the deterioration of once promising fairy-tales, and I also thought that it was necessary to develop a new philosophy or a new religion to give substance to this unsystematic conjecture.

I now realize that these considerations were just another example of intellectualistic conceit and folly. It is conceited to assume that one has solutions for people whose lives one does not share and whose problems one does not know. It is foolish to assume that such an exercise in distant humanitarianism will have effects pleasing to the people concerned. From the very beginning of Western Rationalism intellectuals have regarded themselves as teachers, the world as a school and 'people' as obedient pupils. In Plato this is very clear. The same phenomenon occurs among Christians, Rationalists, Fascists, Marxists. Marxists did not try to learn from those they wanted to liberate; they attacked each other about interpretations, viewpoints, evidence and took it for granted that the resulting intellectual hash would make fine food for the natives (Bakunin was aware of the doctrinarian tendencies of contemporary Marxism and he intended to return all power – power over ideas included – to the people immediately concerned). My own view differed from those just mentioned but it was still a *view*, an abstract fancy I had invented and now tried to sell without having shared even an ounce of the lives of the receivers. This I now regard as insufferable conceit. So – what remains?

Two things remain. I could follow my own advice to address and try to influence only those people whom I think I understand on a personal basis. This includes some of my friends; it may include philosophers I have not met but who seem to be interested in similar problems and who are not too upset by my style and my general approach. It may also include people from different cultures who are attracted, even fascinated by Western science and Western intellectual life, who have started participating in it but who still remember, in thought as well as in feeling the life of the culture they left behind. My account might lessen the emotional tension they are liable to feel and make them see a way of uniting, rather than opposing to each other, the various stages of their lives.

Another possibility is a change of subject. I started my career as a student of acting, theatre production and singing at the Institute for the Methodological Reformation of the German Theatre in the German Democratic Republic. This appealed to my intellectualism and my dramatic propensities. My intellectualism told me that problems had to be solved by thought. My dramatic propensities made me think that hamming it up was better than going through an abstract argument. There is of course no conflict here, for argument without illustration leads away from the human elements which affect the most abstract problems. The arts, as I see them today, are not a domain separated from abstract thought, but complementary to it and needed to fully realize its potential. Examining this function of the arts and trying to establish a mode of research that unites their power with that of science and religion seems to be a fascinating enterprise, and one to which I might devote a year (or two, or three . . .).

Postscript on Relativism

In a critical notice of my book *Farewell to Reason* Andrew Lugg suggests 'that Feyerabend and likeminded social critics should treat relativism with the disdain that they normally reserve for rationalism'.[1] This I have now done, in *Three Dialogues of Knowledge*,[2] where I say that relativism gives an excellent account of the relation between dogmatic world-views but is only a first step towards an understanding of live traditions, and in *Beyond Reason: Essays on the Philosophy of Paul K. Feyerabend*, where I write that 'relativism is as much of a chimaera as absolutism [the idea that there exists an objective truth], its cantankerous twin'.[3] In the same book I call my earlier advice to keep hands off traditions an 'idiocy'.[4] In both cases I raise objections against relativism, indicate why I changed my mind and mention some of the remaining difficulties.

Andrew Lugg adds that my 'commitment to relativism as a general theory (or principled outlook) is considerably less than total and [that I] can plausibly be read as arguing that the trouble with traditional versions of relativism is that they are pitched at too high a level of abstraction'.[5] This is certainly true of what I say in *Farewell* – but anticipations (which I notice only now, as a result of Lugg's comments) occur already in *Science in a Free Society*.[6] There I distinguish between participants and external observers of traditions, describe objectivism as an illusion created by the special position of the former and summarize my arguments in a series of theses, all of them printed in italics. Thesis i reads: *Traditions are neither good nor bad, they simply are*. Thesis ii: *A tradition assumes desirable or undesirable properties only when compared with some tradition*, i.e. only when viewed by participants who see the world in terms of their own values. And so on. This sounds like Protagoras, and I say so, in thesis iii. However, I then describe (theses v and vi) how traditions interact. I discuss two possibilities, a guided exchange and an open exchange. A

---

1. *Can. Journal of Philosophy*, Vol. 21, 1991, p. 116 – received 1989.
2. Oxford, 1991, pp. 151ff. (MS finished 1989/90.)
3. Dodrecht, 1991, p. 515. (MS finished 1989.)
4. Ibid., p. 509.
5. Ibid.
6. London, 1978, part 1, section 2, pp. 27ff – reprinted without change in Chapter 17 of the second edition of *Against Method*, London, 1988, and with added comment in Chapter 17, pp. 225ff of the present edition.

guided exchange adopts 'a well-specified tradition and accept[s] only those responses that correspond to its standards. If one party has not yet become a participant . . . he will be badgered, persuaded, "educated" until he does – and then the exchange begins.' 'A *rational debate*', I continue, 'is a special case of a guided exchange.' In the case of an open exchange 'the participants get immersed into each other's ways of thinking, feeling, perceiving to such an extent that their ideas, perceptions, world-views may be entirely changed – they become different people participating in a new and different tradition. An open exchange respects the partner whether he is an individual or an entire culture, while a rational exchange promises respect only within the framework of a rational debate. An open exchange has no organon though it may invent one; there is no logic though new forms of logic may emerge in its course.' In sum, an open exchange is part of an as yet unspecified and unspecifiable practice.

These comments imply, first, that traditions are rarely well defined (open exchanges are going on all the time) and, secondly, that their interactions cannot be understood in general terms. Keeping traditions alive in the face of external influences, we act in an only partly conscious way. We can describe results after they have occurred, we cannot incorporate them into a lasting theoretical structure (such as relativism). In other words, there cannot be any *theory* of knowledge (except as part of a special and fairly stable tradition), there can at most be a (rather incomplete) *history* of the ways in which knowledge has changed in the past. In my next book I shall discuss some episodes of such a history.

In the meantime I have started using the term 'relativism' again, but in a new sense. In the second edition of the present book I explained this sense by saying that 'Scientists [and, for that matter, all members of relatively uniform cultures] are sculptors of reality.'[7] That sounds like the strong programme of the sociology of science except that sculptors are restricted by the properties of the material they use. Similarly individuals, professional groups, cultures can create a wide variety of surroundings, or 'realities' – but not all approaches succeed: some cultures thrive, others linger for a while and then decay. Even an 'objective' enterprise like science which apparently reveals Nature As She Is In Herself intervenes, eliminates, enlarges, produces and codifies the results in a severely

---

7. Op. cit., p. 270. See also the more detailed account in 'Realism and the Historicity of Knowledge', *Journal of Philosophy*, 1989.

standardized way – but again there is no guarantee that the results will congeal into a unified world. Thus all we *apprehend* when experimenting, or interfering in less systematic ways, or simply living as part of a well-developed culture is how what surrounds us *responds* to our actions (thoughts, observations, etc.); *we do not apprehend these surroundings themselves*: Culture and Nature (or Being, to use a more general term) are always entangled in a fashion that can be explored only by entering into further and even more complicated entanglements.

Now, considering that scientists use different and often contradictory methods of research (I describe some of them in Chapter 19 of the present edition), that most of these methods are successful and that numerous non-scientific ways of life not only survived but protected and enriched their inhabitants we have to conclude that Being responds differently, *and positively*, to many different approaches. Being is like a person who shows a friendly face to a friendly visitor, becomes angry at an angry gesture, remains unmoved by a bore without giving any hint as to the principles that make Him (Her? It? Them?) act the way they do in the different circumstances. What we find when living, experimenting, doing research is therefore not a single scenario called 'the world' or 'being' or 'reality' but a variety of responses, each of them constituting a special (and not always well-defined) reality for those who have called it forth. This is relativism because the type of reality encountered depends on the approach taken. However, it differs from the philosophical doctrine by admitting failure: not every approach succeeds. In my reply to critics[8] I called this form of relativism 'cosmological' relativism, in an article published in *Iride*[9] I spoke of an 'ontological' relativism, in 'Nature as a Work of Art'[10] I argued that the world of modern science (and not only the description of this world) is an artwork constructed by generations of artisan/scientists while in 'Realism and the Historicity of Knowledge'[11] I indicated how such views are related to the ideas of Niels Bohr. In the last article I also mentioned that ontological relativism might be similar to Thomas Kuhn's more recent philosophy.

---

8. In Gonzalo Munevar (ed.), *Beyond Reason*, Dodrecht-Boston-London, 1991, p. 570.

9. No. 8, n.s., January–April 1992.

10. *Common Knowledge*, Vol. 1, No. 3, 1993.

11. op. cit., footnote 7 above.

Having before me a copy of Kuhn's Robert and Maurine Rothschild Distinguished Lecture of 19 November 1991, I can now describe the similarities and the differences in greater detail.

We both oppose the strong programme in the sociology of science. As a matter of fact I would say, exactly as Kuhn does, that 'the claims of the strong programme' are 'absurd: an example of deconstruction gone mad'. I also agree that it is not enough to undermine the authority of the sciences by historical arguments: Why should the authority of history be greater than that of, say, physics? All we can show historically is that a *general* appeal to scientific authority runs into contradictions. That undermines any such appeal; however, it does not tell us how science should now be interpreted or used. (Such questions, I would say, have to be answered by the interested parties themselves, according to their standards, conceptions, cultural commitments).

Kuhn says that 'the difficulties that have seemed to undermine the authority of science should not be simply seen as observed facts about its practice. Rather they are necessary characteristics of any developmental or evolutionary process.' But how do we know that science is an evolutionary process rather than a static way of finding more facts and better laws? Either from 'observed facts about its practice' or from interpretations that are imposed from the outside. In the first case we are back at the situation Kuhn wants to overcome while the second case means that science is being incorporated into a wider (cultural) context – a context that values developments – and is interpreted accordingly (the procedure I mentioned in parentheses above). It seems that is what Kuhn really wants, i.e. he wants to settle the question philosophically, not by appealing to facts. I would agree if I knew that for him this is one way among many and not the only possible procedure.

Summarizing his argument Kuhn makes three assertions. 'First, the Archimedian platform, outside history, outside of time and space, is gone beyond recall.' Yes, and no. It is gone as a structure that can be described and yet shown to be independent of any description. It is not gone as an unknown background of our existence which affects us, but in a way which forever hides its essence. Nor is Archimedianism gone as a possible approach. It would be the politically correct approach in a theocracy, for example.

Secondly, Kuhn says that in the absence of an Archimedian platform 'comparative evaluation is all there is'. That is of course true – and trivially so. Thirdly, he challenges the traditional notion of truth as correspondence to reality. 'I am not suggesting, let me emphasize, that there is a reality which science fails to get at. My point is, rather, that no sense can be made of the notion of a reality as it has ordinarily functioned in the philosophy of science.' Here I agree with the proviso that more metaphysical notions of reality (such as those proposed by Pseudo Dionysius Areopagita) have not yet been disposed of.

Let me repeat that the cultures that call forth a certain reality and these realities themselves are never well defined. Cultures change, they interact with other cultures and the indefiniteness resulting therefrom is reflected in their worlds. This is what makes intercultural understanding and scientific change possible: potentially every culture is all cultures. We can of course imagine a world where cultures are well defined and strictly separated and where scientific terms have finally been nailed down. In such a world only miracles or revelation could reform our cosmology.

Index